T0176958

Wearable Computing

Wearable Computing

From Modeling to Implementation of
Wearable Systems Based on Body
Sensor Networks

Giancarlo Fortino, Raffaele Gravina, and Stefano Galzarano

University of Calabria
Rende, Italy

Registered Office
John Wiley & Sons, Inc., 111 River Street, Hoboken, NJ 07030, USA

Editorial Office
111 River Street, Hoboken, NJ 07030, USA

For details of our global editorial offices, customer services, and more information about Wiley products visit us at www.wiley.com.

Wiley also publishes its books in a variety of electronic formats and by print-on-demand. Some content that appears in standard print versions of this book may not be available in other formats.

Library of Congress Cataloging-in-Publication Data
Names: Fortino, Giancarlo, 1971– author. | Gravina, Raffaele, 1982– author. |
   Galzarano, Stefano, 1984– author.
Title: Wearable computing : from modeling to implementation of wearable systems based on
   body sensor networks / Giancarlo Fortino, Raffaele Gravina, Stefano Galzarano.
Description: 1st edition. | Hoboken, NJ : John Wiley & Sons, 2018. | Includes bibliographical
   references and index. |
Identifiers: LCCN 2017053912 (print) | LCCN 2017059016 (ebook) |
   ISBN 9781119078821 (pdf) | ISBN 9781119078838 (epub) | ISBN 9781118864579 (cloth)
Subjects: LCSH: Wearable computers. | Sensor networks.
Classification: LCC QA76.592 (ebook) | LCC QA76.592 .F67 2018 (print) | DDC 004.167–dc23
LC record available at https://lccn.loc.gov/2017053912

Cover design by Wiley
Cover images: © nopporn/Shutterstock; © Sergey Nivens/Shutterstock

Set in 10/12pt Warnock by SPi Global, Pondicherry, India

Printed in the United States of America

10 9 8 7 6 5 4 3 2 1

Contents

Preface

Wearable computing is a relatively new area of research and development that aims at supporting people in different application domains: health care, fitness, social interactions, video games, and smart factory. Wearable computing is based on wearable sensor devices (e.g. to measure heart rate, temperature, or blood oxygen), common life objects (e.g. watch, belt, or shoes), and personal handheld devices (e.g. smartphones or tablets). Wearable computing has been recently boosted by the introduction of body sensor networks (BSNs), i.e. networks of wireless wearable sensor nodes coordinated by more capable coordinators (smartphones, tablets, and PCs).

In particular, BSNs enable a very wide range of application scenarios in different industry sectors. We can categorize them into different domains: e-Health, e-Emergency, e-Entertainment, e-Sport, e-Factory, and e-Social.

e-Health applications span from early detection or prevention of diseases, elderly assistance at home, to post-trauma rehabilitation after surgeries. e-Emergency applications include BSN systems to support fire fighters, response teams in large-scale disasters due to earthquakes, landslides, terrorist attacks, etc. e-Entertainment domain refers to human–computer interaction systems typically based on BSNs for real-time motion and gesture recognition. e-Sport applications are related to the e-Health domain, although they have a non-medical focus. Specifically, this domain includes personal e-fitness applications for amateur and professional athletes, as well as enterprise systems for fitness clubs and sport teams offering advanced performance monitoring services for their athletes. e-Factory is an emerging and very promising domain involving industrial process management and monitoring, and workers' safety and collaboration support. Finally, e-Social applications may use BSN technologies to recognize user emotions and cognitive states to enable new forms of social interactions with friends and colleagues. An interesting example is given by a system that involves the interaction between two people's BSNs to detect handshakes and, subsequently monitor their social and emotional interactions.

Although the basic elements (sensors, protocols, and coordinators) of a BSN are available (already from a commercial point of view), developing BSN systems/applications is a complex task that requires design methods based on effective and efficient programming frameworks. In this book, we will provide programming approaches and methods to effectively develop efficient BSN systems/applications. Moreover, we also provide new techniques to integrate BSN-based wearable systems with more general Wireless Sensor Network systems and with Cloud computing.

This book, entitled *Wearable Computing: From Modeling to Implementation of Wearable Systems Based on Body Sensor Networks*, is based on an intense and extensive basic and applied research activity driven by the SPINE project (http://spine.deis.unical.it), whose authors are cofounders, responsible, and main developers. Thus, the book is connected to the SPINE website to provide readers with software and tools for the development of their wearable computing systems.

This book is aimed at a large audience in the Wearable Computing domain, that is gaining considerable research interest and momentum, and is expected to be of increasing interest to academic researchers and particularly to commercial developers. Upon reading this book the audiences will perceive the following benefits:

- Learn the state-of-the-art in research and development on wearable computing, wireless BSNs, wearable systems integrated with mobile computing, wireless networking, and cloud computing.
- Obtain a future roadmap by learning advanced technology and open research issues.
- Gather the background knowledge to tackle key problems, whose solutions will enhance the evolution of next-generation wearable systems.
- Use the book as a valuable reference for a technical professional in a related industry.
- Use the book as a text book in the late undergraduate or the graduate level to prepare students who intend to perform research in the field of the book or intend to be employed in a related industry.

The main topics of the book are the following:

- *Wearable Computing*, the study or practice of inventing, designing, building, or using miniature body-borne computational and sensory devices. Wearable computers may be worn under, over, or in clothing, or may also be themselves clothes.
- *Wireless Sensor Networks (WSNs)*, collections of tiny devices capable of sensing, computation, and wireless communication operating in a certain environment to monitor and control events of interest in a distributed manner and collectively react to critical situations. WSN applications span various domains such as environmental and building monitoring and surveillance,

pollution monitoring, agriculture, health care, home-automation, energy management, earthquake, and eruption monitoring.

- *Body Sensor Networks (BSNs)*, involving wireless wearable physiological sensors applied to the human body for medical and nonmedical purposes. In particular, they allow for the continuous measurement of body movements and physiological parameters, such as heart rate, muscular tension, skin conductivity, and breathing rate and volume, during the daily life of a user.
- *In-node Signal Processing*, a central computing method in advanced wireless sensor platforms through which data processing is carried out directly on the sensor node to preprocess data acquired from sensors, to fuse data coming from other sensor nodes, and, notably, to perform higher level computation such as classification and decision making.
- *Mobile Computing*, human–computer interaction by which a computer is expected to be transported during normal usage. Mobile computing involves mobile communication, mobile hardware, and mobile software. Communication issues include ad-hoc and infrastructure networks as well as communication properties, protocols, data formats, and concrete technologies. Hardware includes mobile devices or device components. Mobile software deals with the characteristics and requirements of mobile applications.
- *Cloud Computing*, the use of computing resources (hardware and software) that are delivered as a service over a network (typically the Internet). The name comes from the use of a cloud-shaped symbol as an abstraction for the complex infrastructure it contains in system diagrams. Cloud computing entrusts remote services with a user's data, software, and computation.
- *Platform-Based Design (PBD)*, an embedded computing design methodology that consists of a sequence of design/development steps that leads the initial high-level description of a digital system to its final implementation. Each step is a refinement process that transforms the design from a higher level description to a lower level description that is progressively closer to the final implementation.
- *Software Framework*, an abstraction in which software providing generic functionality can be selectively changed by user code, thus providing application-specific software. A software framework is a universal, reusable software platform used to develop applications, products, and solutions. Software Frameworks include support programs, compilers, code libraries, an application programming interface (API), and tool sets that bring together all the different components to enable development of a project or solution.
- *Autonomic Computing* is a paradigm born as a response to the increasing complexity of managing computing systems. It faces the problem by introducing a series of self-* properties (self-configuration, self-healing, self-optimization, and self-protection) into complex systems, through which such systems can be capable of performing several self-management actions without any human intervention.

- *Activity Recognition* aims to recognize the actions and goals of one or more agents from a series of observations on the agents' actions and the environmental conditions. Since the 1980s, this research field has captured the attention of several computer science communities due to its strength in providing personalized support for many different applications and its connection to many different fields of study such as medicine, human–computer interaction, or sociology. Specifically, we are mainly interested in sensor-based single-user and multiuser activity recognition that integrates the emerging area of sensor networks with novel data mining and machine learning techniques to model a wide range of human activities.

Specifically, this book is organized into 12 chapters:

- Chapter 1, Body Sensor Networks (BSNs), covers the state-of-the-art about wearable sensor nodes, network architecture/protocols/standards, and applications/systems.
- Chapter 2, BSN Programming Frameworks, analyzes the state-of-the-art about the most known software frameworks (CodeBlue, Titan, RehabSPOT, and others) for programming BSN applications/systems.
- Chapter 3, Signal Processing In-Node Environment, describes in detail the SPINE framework (http://spine.deis.unical.it) from architectural and programming perspectives.
- Chapter 4, Task-Oriented Programming, discusses task-oriented programming of BSN applications through SPINE2.
- Chapter 5, Autonomic BSNs, illustrates how to make BSNs autonomic, by using SPINE*, an extension of SPINE2.
- Chapter 6, Agent-oriented BSNs, presents the use of the Agent paradigm for programming BSN systems. Specifically, the MAPS (Mobile Agent Platform for SunSPOT) framework is used to design and implement agent-based BSNs.
- Chapter 7, Collaborative BSNs, provides an introduction of methods and architectures to make BSNs interact with each other for supporting multiuser BSN applications.
- Chapter 8, Integration of BSNs and Wireless Sensor Networks, covers gateway-based solution for interoperability between BSNs and infrastructural WSNs (e.g. building indoor sensor networks). This would enable "invisible" interaction between BSN-worn people and the surrounding environment.
- Chapter 9, Integration of Wearable and Cloud Computing, presents an architecture for the integration of BSNs and the Cloud, called BodyCloud, based on Google App Engine. It is crucial now to move the data acquired or preprocessed on the human body to the cloud for storing and nonreal-time analysis purposes.
- Chapter 10, Development Methodology for BSN Systems, describes a SPINE-based methodology for the development of BSN systems.

The methodology guides the BSN system developer from requirement analysis to implementation and deployment.

- Chapter 11, SPINE-based BSN Applications, presents several applications developed through SPINE in different application domains (Activity Recognition: recognition of human postures and movements, Emotion Recognition: recognition of stress and fear, Handshake Detection: collaborative recognition of two people's handshake, and Rehabilitation: real-time computation of extension angles of elbow/knee).
- Chapter 12, SPINE at Work, provides a quick yet effective reference for BSN programmers interested in developing their applications using the SPINE framework. The chapter provides the necessary information for setting up the SPINE environment so as to start programming as well as insights on how the framework itself can be customized and extended.

Acknowledgments

This book is the result of direct and indirect involvement of many researchers, academics, and industry professionals.

We sincerely thank all the other members of the SPINE team: Fabio Bellifemine, Roberta Giannantonio, Antonio Guerrieri, Roozbeh Jafari, and Alessia Salmeri. Our gratitude also goes to all the international researchers and internal alumni that contributed to the SPINE Project with studies, programming efforts, and novel ideas; in particular let us remind Andrea Caligiuri, Giuseppe Cristofaro, Philip Kuryloski, Vitali Loseu, Ville-Pekka Seppa, Edmund Seto, Marco Sgroi, and Filippo Tempia.

This work has been partially carried out under the framework of INTER-IoT, Research and Innovation action – Horizon 2020 European Project, Grant Agreement 687283, financed by the European Union.

We thank Wiley's publication staff for handling the book project and supporting its publication.

We hope that this book will serve as a valuable text for academic researchers and particularly to commercial developers working in the wearable computing domain.

1

Body Sensor Networks

1.1 Introduction

This chapter provides an overview of the state-of-the-art and technology in the field of wireless body sensor networks (BSNs). After introducing the motivations and the potential applications of this emerging technology, the chapter focuses on the analysis of the architecture of sensor nodes, communication techniques, and energy issues. We will then present and compare some of the programmable sensing platforms that are most commonly used in the context of wireless sensor networks (WSNs), and in particular those applied to remote monitoring of patients. The chapter also contains an analysis of relevant vital human signals and physical sensors used for their recording. Finally, the chapter presents the hardware/software characteristics that must be taken into consideration during the design stages of a healthcare monitoring system based on BSNs. For instance, important characteristics are sensor wearability, biocompatibility, energy consumption, security, and privacy of the acquired biophysical information.

1.2 Background

The widespread use of mobile applications for patient monitoring over the last few years is radically changing the approach to the health care. In today's society, this is gaining an increasingly important role in the prevention of diseases; the convenience, for instance in terms of health-care costs, is significant. The BSN technology makes often use of mobile applications that allow for the transmission to a coordinator node, such as a smartphone or a tablet, information about vital signs and physical activities (movements and gestures) [1, 2]. The miniaturization and the production cost reduction are leading to the realization of extremely small-sized sensing and computing devices with high processing capacity thus giving a great impulse to the development of WSNs, and, as a

Wearable Computing: From Modeling to Implementation of Wearable Systems Based on Body Sensor Networks, First Edition. Giancarlo Fortino, Raffaele Gravina, and Stefano Galzarano.
© 2018 John Wiley & Sons, Inc. Published 2018 by John Wiley & Son, Inc.

direct consequence, of BSNs. Very heterogeneous information and diversified physiological signals can be transmitted, possibly after the application of sensor fusion techniques [3], by the sensor nodes to the coordinator device.

Figure 1.1 shows a number of wearable sensing devices and their typical location on the body:

1) *Electrocardiography* (ECG): the ECG is used to record the electrical activity (including the heart rate) of the heart over a period of time using electrodes placed on the skin.
2) *Blood pressure meter*: also known as sphygmomanometer, it is a device used to measure (typically, both diastolic and systolic) blood pressure.

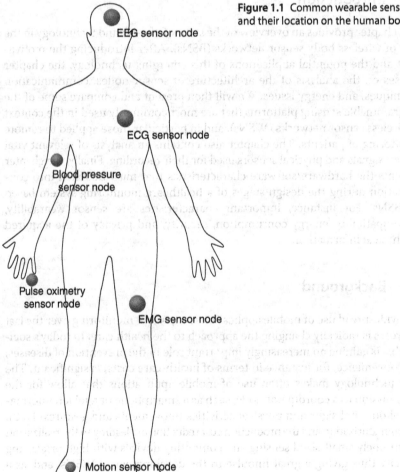

Figure 1.1 Common wearable sensors and their location on the human body.

EEG sensor node

ECG sensor node

Blood pressure sensor node

Pulse oximetry sensor node

EMG sensor node

Motion sensor node

3) *Pulse oximetry*: the oximeter is a medical device that allows us to measure noninvasively the amount of hemoglobin in the blood. Since hemoglobin binds with oxygen, it is therefore possible to obtain an estimate of the amount of oxygen present in the blood.

4) *Electromyography* (EMG): the EMG sensor is used to monitor muscle activity, using a needle electrode inserted into the muscle for high accuracy, or, more practical and noninvasive, with simple skin electrodes. It records the activity of the muscle fibers under different conditions: at rest, during voluntary contraction up to the maximum effort, and during a sustained average contraction.

5) *Electroencephalography* (EEG): the EEG sensor uses electrodes placed on the scalp to monitor the brain activity and capture different types of brain waves.

6) *Motion* inertial sensors (e.g. accelerometers and gyroscopes) monitor human movements and even gestures.

BSN systems are commonly characterized by a number of hardware and software requirements:

1) *Interoperability*: it is necessary to ensure the continuous data transfer through different standards (e.g. Bluetooth and ZigBee) to promote the exchange of information and ensure interaction between devices. In addition, it should provide an adequate level of scalability in relation to the number of sensor nodes and the workload of the BSN.

2) *System device*: the sensors must be of low complexity, small size, lightweight, energy efficient, easy to use, and reconfigurable. In addition, patient biosignal storage, retrieval, visualization, and analysis must be facilitated.

3) *Security* at the device and system level: particular attention must be paid to secure transmission and authenticated access to such sensible data.

4) *Privacy*: the BSN could be considered as a "threat" to the freedom of the individual, if the purpose of the applications goes "beyond" the medical purposes. Social acceptance to these systems is the key to their wider dissemination.

5) *Reliability*: the whole system must be reliable at hardware, network, and software levels. Reliability affects directly the quality of monitoring because, in the worst case, the failure to observe and/or successfully notify a "critical risk event" can be lethal for the patient. Because of the limitations and requirements on communication and power consumption, the reliability techniques used in traditional networks are not easily applicable in the BSN domain and, both at the design and implementation phase, this must be taken seriously.

6) *Validation and accuracy* of sensory data: sensing devices are subject to hardware constraints that can affect the quality of the acquired data; both wired and wireless connections are not always reliable; environmental

interference and limited energy availability also affect this aspect. This can cause inconsistencies in the transmitted data and might lead to critical errors in their interpretation. It is very important that all data transmitted from the sensor nodes to the coordinator are adequately "validated" either in hardware or software, trying to identify the "critical points" of the system.

7) *Data consistency*: for large-scale BSNs, with many and heterogeneous sensors, a single biophysical phenomenon may be "fragmented" and only partially detectable into individual signals. This aspect arises problems of information consistency, which must be addressed through appropriate synchronization strategies, data fusion techniques [3], and/or mutual exclusion in the access to data.

8) *Interference*: wireless links used in the BSN should try to minimize the interference issues and favor the coexistence of sensor nodes with other network devices available within the radio range.

9) *Biological compatibility*: the wearable sensors and skin electrodes must be biocompatible and stable, as they might operate on the user for a long period of time without interruptions.

In addition to the hardware and software features, we highlight some aspects that could encourage the wide diffusion and exploitation of BSN systems:

1) *Costs*: users expect low costs for health monitoring, yet preserving high performance of the devices used.

2) *Different levels of monitoring*: users may require different levels of monitoring, for example, to control the risk of ischemic heart disease or of falling during movements. Depending on the operating mode, the energy level required for the power supply of the devices can also vary.

3) *Noninvasive easy-to-use devices*: the devices must be wearable, lightweight, and noninvasive. They should not hinder users in their daily activities; their operation must be "transparent" to users who should ignore the details of the monitoring task.

4) *Consistent performance*: sensors must be calibrated and accurate, and they should provide consistent measurements even if the BSN is stopped and restarted several times. Wireless links should be as robust as possible and be able to operate correctly in different (noisy) working environments.

1.3 Typical m-Health System Architecture

Figure 1.2 shows the typical architecture of an m-Health system based on BSN technology. It usually consists of three different tiers communicating through wireless (or sometimes wired) channels [4].

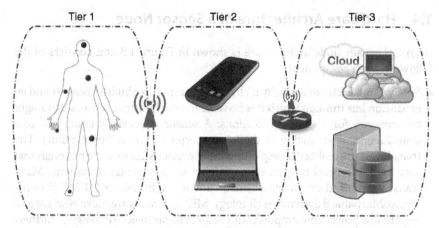

Figure 1.2 A three-tier hierarchical BSN architecture: (1) body sensor tier, (2) personal area network tier, and (3) global network tier.

Tier 1 represents the *Body Sensor Tier* and includes a set of wireless wearable medical sensor nodes composing the BSN. Each node is able to detect, sample, and process one or more physiological signals. For example, a motion sensor for discriminating postures, gestures, and activities; an electrocardiogram (ECG) sensor can be used for monitoring cardiac activity; and an electroencephalogram (EEG) sensor for monitoring cerebral electrical activity, and so on.

Tier 2 is the *Personal Area Network Tier* and contains the personal coordinator device (often a smartphone or a tablet, but possibly a PC) running an end-user application. This tier is responsible for a number of functions providing a transparent interface to the BSN, to the user, and to the upper tier. The interface to the BSN provides functionalities to configure and manage the network, such as sensor discovery and activation, sensory data recording and processing, and establishment of a secure communication with both Tier 1 and Tier 3. When the BSN has been configured, the end-user monitoring application starts providing feedback through a user-friendly graphical and/or audio interface. Finally, if there is an active channel of communication with the upper tier, it can report raw and processed data for off-line analysis and long-term storage. Conversely, if Internet connectivity is temporary unavailable, the coordinator device should be able to store the data locally and perform the data transfer as soon as the connectivity is restored.

Tier 3 is the *Global Network Tier* and comprises one or more remote medical servers or a Cloud computing platform. Tier 3 usually provides services to medical personnel for off-line analysis of a patient's health status, real-time notification of life-critical events and abnormal conditions, and scientific and medical visualization of collected data. In addition, this tier can provide a web interface for the patient itself and/or relatives too.

1.4   Hardware Architecture of a Sensor Node

A typical sensor node architecture is shown in Figure 1.3 and consists of the following main components:

- *Sensing* unit, each node usually includes one or multiple built-in sensors and an expansion bus through which it is possible to attach further sensors that might be necessary for specific applications. A sensor is generally composed of a transducer and an analog-to-digital converter (see next bullet point). The transducers are realized by exploiting the characteristics of some materials that vary their "electrical properties" to varying environmental conditions. Many transducers used on wireless sensor nodes are based on MEMS (Micro-ElectroMechanical Systems) technology. MEMS sensors are more efficient and require less power consumption with respect to piezoelectric sensors; furthermore, MEMS sensors are characterized by low production costs, although this could lead to less precision if compared with piezoelectric sensors.
- *Analog-to-Digital Converter* (ADC) converts the voltage value of a transducer into a digital value, which will then be used for post-processing.
- *Processing* unit, the Micro-Controller Unit (MCU) of a sensor node is usually associated with a built-in limited memory unit to improve the processing speed and enable local online sensory data processing. The sensor node is, therefore, able to perform signal processing such as "background noise" filtering, data fusion and aggregation, and feature extraction (e.g. mean, variance, maximum/minimum value, entropy, and signal amplitude/energy). The MCU is also responsible for the management of the other hardware resources.
- *Transceiver* unit is the component that connects the node to the network. It can be an optical or a radio frequency (RF) device. It is also possible, and actually very useful, to use the radio with a low duty-cycle, to help reducing the power consumption.

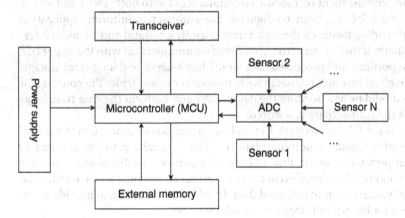

Figure 1.3 Typical hardware architecture of a sensor node.

- *External memory* is needed to store the binary code of the program running on the sensor node. Some sensor platforms also include a further memory (usually a microSD flash memory) as a mass storage unit for sensory data recording.
- *Power supply* is the scarcest resource of a sensor node and must be preserved as much as possible to prolong its lifetime; it could be notably supported by a unit for energy harvesting (e.g. from solar light, heat, or vibration).

1.5 Communication Medium

In a multi-hop sensor network the nodes can interact with each other via a wireless communication medium. One choice is to use the ISM (industrial, scientific, and medical) radio spectrum [5], i.e. a predefined set of frequency bands that can be used freely in many countries. Most of the sensors currently on the market do in fact make use of a RF circuit. Another option is given by infrared (IR) communication. On the one hand, the IR communication does not require permits or licenses, it is protected from interference, and IR transceivers are very cheap and easy to realize. On the other hand, however, IR requires line-of-sight between the transmitter and the receiver, which makes it hardly usable for WSNs and BSNs as nodes very often cannot be deployed in such a way.

1.6 Power Consumption Considerations

A sensor node is normally equipped with a very limited energy source. The lifecycle of a sensor node heavily depends on the battery dimensions and on the processing and communication duty-cycling. For these reasons, many research efforts are focusing on the design of power-aware communication protocols and algorithms, with the aim of optimizing energy consumption. While in traditional mobile networks and ad-hoc networks energy consumption is not the most important constraint, in the WSN domain it is a crucial aspect. This is true even in the specific subdomain of the BSNs. Although it is generally easier to recharge or replace the batteries of the wearable nodes, due to wearability reasons, the battery dimension (and hence its capacity) is generally much smaller than in other WSN scenarios.

In a sensor node, the energy consumption is mainly due to three tasks:

- *Communication*: it is the most affecting factor. Low-power radios, strict radio duty-cycling, power-aware WSN-specific communication protocols and standards, and on-node data fusion and aggregation techniques are critical design choices for reducing the activation of the transceiver module as much as possible. It is worth noting that both transmission and listening/reception time must be optimized.

- *Sensing*: the power required to carry out the sampling depends on the nature of the application and, as a consequence, on the type of the physical transducers involved.
- *Data processing*: it must be taken into account, even though the energy consumed for processing a given amount of data is very small compared to the energy requirements for transmitting the same amount of data. Experimental studies showed that the energy cost for transmitting 1 kB of data is about the same that would be obtained by performing 3–100 million instructions on the sensor node microcontroller [6].

1.7 Communication Standards

The aforementioned requirements impose very tight restrictions on the type of network protocols that can be used in WSNs. The short-range wireless technologies are a prerequisite, given the limited power budget available for each node. The implementation of a wireless network communication protocol that must be robust, fault tolerant, and capable of self-configuration even in hostile environments represents a considerable technological challenge, which required (and still requires) the efforts of several standardization bodies, such as IEEE and IETF.

The **IEEE 802.15.4** [7] is to date the most widely adopted standard in the WSN domain. Indeed, it is intended to offer the fundamental lower network layers (physical and MAC) of Wireless Personal Area Networks (WPANs) focusing on low-cost, low-speed ubiquitous communication between devices. The emphasis is on very low-cost communication of nearby devices with little to no underlying infrastructure. The basic protocol conceives a 10 m communication range with a transfer rate of 250 kbit/s. Tradeoffs are possible to favor more radically embedded devices with even lower power requirements, through the definition of several physical layers. Lower transfer rates of 20 and 40 kbit/s were initially defined, with the 100 kbit/s rate being added later. Even lower rates can be considered with the resulting effect on power consumption. The main identifying feature of 802.15.4 is the importance of achieving extremely low manufacturing and operation costs, and technological simplicity, without sacrificing flexibility or generality. Important features include real-time suitability by reservation of guaranteed time slots, collision avoidance through CSMA/CA, and integrated support for secure communications. It operates on one of three possible unlicensed frequency bands:

- 868.0–868.6 MHz: Europe, allows 1 communication channel.
- 902–928 MHz: North America, up to 30 channels.
- 2400–2483.5 MHz: Worldwide use, up to 16 channels.

To complete the IEEE 802.15.4 standard, the **ZigBee** [8] protocol has been realized. ZigBee is a low-cost, low-power, wireless mesh network standard built upon the physical layer and medium access control defined in the 802.15.4. It is intended to be simpler and less expensive than, for instance, Bluetooth. ZigBee chip vendors typically sell integrated radios and microcontrollers with 60 to 256 kB flash memory. The ZigBee network layer natively supports both star and tree networks, and generic mesh networks. Every network must have one coordinator device. In particular, within star networks, the coordinator must be the central node. Specifically, the ZigBee specification completes the 802.15.4 standard by adding four main components:

- *Network layer*, which enables the correct use of the MAC sublayer and provides a suitable interface for the application layer.
- *Application layer* is the highest-level layer defined by ZigBee and represents the interface to the end users.
- *ZigBee device object* (ZDO) is the protocol responsible for overall device management, security keys, and policies. It is responsible for defining the role of a device (i.e. coordinator or end device).
- *Manufacturer-defined application objects*, which allow for customization and favor total integration.

Bluetooth [9] is a proprietary open wireless technology standard for exchanging data over short distances (using short wavelength radio transmissions in the ISM band from 2400 to 2480 MHz) from fixed and mobile devices, creating WPANs with high levels of security. Bluetooth uses a radio technology called frequency-hopping spread spectrum, splitting the data being sent into portions and transmitting the portions on up to 79 bands (1 MHz each). Bluetooth is a packet-based protocol with a master–slave structure. One master may communicate with up to seven slaves in a so-called *piconet*; all devices share the master's clock. Packet exchange is based on the basic clock, defined by the master. The specification also provides for the connection of two or more piconets to form a *scatternet*, in which certain devices simultaneously play the master role in one piconet and the slave role in another. Although being designed for WPANs, the first versions of Bluetooth are actually suitable only for BSN systems that do not require long battery life before recharging. This is because the Bluetooth power consumption profile is significantly higher compared with 802.15.4. Other factors limiting the use of Bluetooth in the BSN domain are the high communication latency (typically around 100 ms) and the long setup time (that, due to the discovery procedure, can take several seconds).

To overcome these limitations, Bluetooth released the 4.0 version that has been called **Bluetooth Low Energy (BLE)** [10]. One of the BLE design driving factors is the specific support for applications such as health care, sport, and fitness. The promoter for such applications is the Bluetooth Special Interest

Group in cooperation with the Continua Health Alliance. BLE operates in the same spectrum range (2400–2480 MHz) as classic Bluetooth but uses a different set of channels. Instead of 79 1-MHz wide channels, BLE uses 40 2-MHz wide channels. BLE is designed with two implementation alternatives: *Single mode* and *dual mode*. Small devices like watches and sport sensors based on a single-mode BLE implementation will take advantage of the low power consumption and low production costs. However, pure BLE is not backward compatible with the classic Bluetooth protocol. In dual-mode implementations, instead, the new low-energy functionality is integrated into classic Bluetooth circuitry. The architecture will share classic Bluetooth technology radio and antenna, enhancing current chips with the new low-energy stack.

ANT [11] is an ultra-low-power wireless communications protocol stack operating in the 2.4 GHz band. A typical ANT protocol transceiver comes preloaded with the protocol software and must be controlled by an application processor. It is characterized by a low computational overhead and high efficiency, resulting in low power consumption by the radios supporting the protocol. Similar to BLE, ANT has been targeted for sport, wellness, and home health monitoring, among other WSN application scenarios. To date, indeed, ANT has been adopted in a number of commercial wrist-mounted instrumentation, heart rate monitoring, speed and distance monitoring, bike computers, and health and wellness monitoring devices.

The **IEEE 802.15 WPAN Task Group 6 (BAN)** [12] is developing a communication standard specifically optimized for low-power devices operating on, in, or around the human body to serve a variety of applications including medical, consumer electronics, personal entertainment, and others. Compared to IEEE 802.15.4, IEEE 802.15.6 focuses specifically on BSNs, addressing their identifying characteristics such as shorter communication range (the standard supports a range of 2–5 m) and larger data rate (up to 10 Mbps), which help in decreasing power consumption and meeting safety and biofriendly requirements.

1.8 Network Topologies

The most common network topologies adopted in the BSN domain are the following:

- peer-to-peer
- star
- mesh
- clustered

The *peer-to-peer* (P2P) topology (see Figure 1.4) reflects BSN systems that do not rely on a coordinator station to operate. It is worth noting that a pure P2P topology is never used in practice today. Even for systems where the sensor

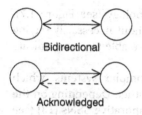

Bidirectional

Acknowledged

Figure 1.4 Peer-to-peer topology.

nodes adopt a decentralized communication paradigm to reach a certain common goal, there is at least one node that interfaces with the user to receive commands and provide some sort of feedback for the events generated by the BSN.

The most common network topology for a BSN system is actually the *star* (see Figure 1.5). Here, the coordinator device acts as the center of the star and it is in charge of configuring the remote sensor nodes (which do not communicate among each other directly), and gathering the sensory information.

The P2P and star topologies are used for personal BSN applications (e.g. health monitoring, wellness, or sport) that do not need to interact with other BSNs.

The *mesh* topology (see Figure 1.6) is an extension of the star, where multiple BSNs may interact, and even collaborate, through the existence of an underlying infrastructure consisting of gateway nodes necessary to enable the communication among BSNs.

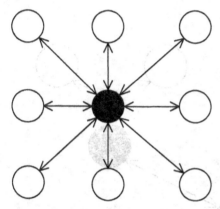

Figure 1.5 Star topology.

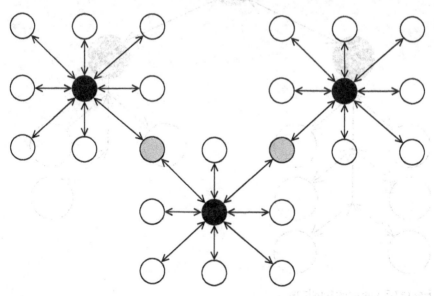

Figure 1.6 Mesh topology.

Somewhat similar to the mesh is the clustered topology (see Figure 1.7). Here, however, different BSNs may communicate without necessarily relying on a fixed infrastructure. In other words, the BSNs are able to communicate directly, typically in a P2P fashion.

Mesh and clustered topologies are adopted in complex systems, which involve different BSNs to communicate among each other. Depending on the specific application, they are often referred to as Collaborative BSNs [13] (see Chapter 7).

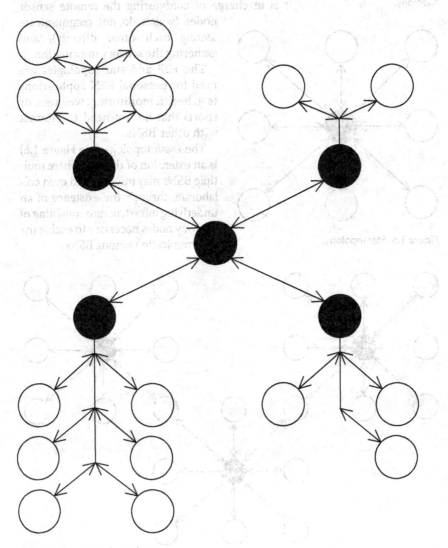

Figure 1.7 Clustered topology.

1.9 Commercial Sensor Node Platforms

A comprehensive analysis on commercial sensor platforms for BSN applications is out of the scope of this section. However, to provide a overview on their current status, a brief list is summarized in Table 1.1. An interesting survey on sensor network platforms can be found in Ref. [14].

In the following, we just briefly describe the main technical specifications of some of the most popular sensor node architectures.

The Intel Mote [15] is among the first wireless sensor node platforms; built on a motherboard of 3×3 cm and equipped with an Intel XScale PXA270 processor with 32 MB of flash memory and 32 MB of SDRAM, it allows for high-performance computing. It integrates an 802.15.4 radio, while additional wireless standards, such as Bluetooth and 802.11b, are supported by means of attachable boards.

The Mica Mote [16] (see Figure 1.8), developed at the University of California at Berkeley, is used for research and development of networks with low-power consumption requirements. It is equipped with an Atmel ATMEGA128 micro-controller at 4–16 MHz (on the MicaZ) with 128 kB of Flash and 4 kB of SRAM. The radio module is based on an RF transmitter at 916.5 MHz on the Mica, while on the CC2420 at 2.4 GHz on the MicaZ. The platform is distinguished by the high number of additional plug-in sensor boards.

The TelosB (also known as Tmote Sky) [17] (see Figure 1.9) is an open-source low-power wireless sensor node platform designed by the University of California, Berkeley, for pervasive monitoring applications and for rapid proto-typing of WSN systems. It integrates an 8 MHz Texas Instruments MSP430 microcontroller, humidity, temperature and light sensors, and an IEEE 802.15.4 compliant Chipcon CC2420 radio module.

The Shimmer nodes [18] (see Figure 1.10) are specifically designed to sup-port wearable medical applications and provide a highly extensible platform, by means of plug-in sensor boards, for real-time detection of movements and changes in physiological parameters. They are among the smaller nodes on the market and have a plastic cover that protects the internal electronics and the battery. Furthermore, the size and the wide availability of elastic straps (e.g. for arms, chest, wrist, waist, and ankle) makes this platform probably the most appropriate for developing BSN-based m-Health systems. Currently, there are four commercial revisions of the platform: Shimmer, Shimmer2, Shimmer2R, and Shimmer3. All of them have the same MCU (TI MSP430) and the same radio chipset (CC2420), support local storage media microSD, are powered by a rechargeable lithium battery, and support Bluetooth communication, thanks to a second dedicated radio module. The Shimmer3 revision is slightly differ-ent as it uses a more powerful 24 MHz MSP430 microcontroller and includes natively only the Bluetooth radio, while offering an expansion interface for connecting an additional radio or a coprocessor. The Bluetooth support is an

Table 1.1 List of commercial sensor node platforms.

Sensor platform	MCU	Transceiver	Code/data memory	External memory	Programming language
BTNode	ATmega 128L 8 MHz	802.15.4 (CC1000), Bluetooth	180/64 kB	128 kB	C, nesC/TinyOS
Epic mote	TI MSP430 8 MHz	802.15.4 (CC2420)	48/10 kB	2 MB Flash	nesC/TinyOS
MicaZ	ATMega 128 16 MHz	802.15.4 (CC2420)	128/4 kB	512 kB	nesC/TinyOS
Shimmer3	TI MSP430 24 MHz	Bluetooth	256/16 kB	2 GB microSD	C, nesC/TinyOS
SunSPOT	ARM920T 180 MHz		512 kB	4 MB Flash	JavaME
TelosB	TI MSP430 8 MHz	802.15.4 (CC2420)	48/10 kB	1 MB Flash	C, nesC/TinyOS
Waspmote	ATMega 1281 8 MHz	ZigBee or Bluetooth or Wifi	128/8 kB	2 GB microSD	C
Intel Mote	XScale PXA270 13–416 MHz	802.15.4 (CC2420), Bluetooth, 802.11b	32 MB/32 MB	—	C, TinyOS

Figure 1.8 Mica Mote.

Figure 1.9 TelosB Tmote Sky.

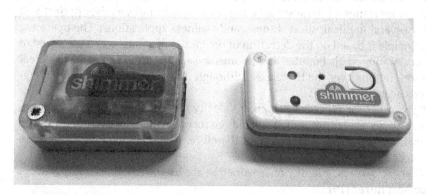

Figure 1.10 Different revisions of the Shimmer platform.

important aspect of this platform as it strengthens the motivation for its use in market-ready m-Health systems, since current smartphones and tablets do have Bluetooth connectivity, but do not support the IEEE 802.15.4 standard.

1.10 Biophysiological Signals and Sensors

There exist several and very different vital signs and biophysiological parameters. Some of them are very useful for realizing effective smart-Health systems. Among the main parameters of interest, there are:

- blood pressure
- blood oxygenation
- blood glucose concentration
- body temperature
- brain activity
- thoracic impedance
- breathing rate
- breathing volume
- cardiac electric activity
- heart rate
- skin conductivity
- muscle activity
- posture and physical activities

There exist wearable noninvasive sensors that can be used to measure, directly or indirectly, each of the aforementioned parameters. One or multiple sensors are typically included in the basic sensor platforms and additional sensors may be integrated through expansion interfaces. In particular, the following physical sensors have been commonly used in research and industrial m-Health systems:

- *Accelerometers* for measuring body movements and gestures. In recent years, the importance of these sensors increased significantly, as they perfectly fit for several medical, sport, fitness, and wellness applications. The operating principle is based on the detection of the inertia of a mass when subjected to acceleration [19]. Popular accelerometer sensors are today able to detect accelerations over the three axes, although there are also two-axis and one-axis accelerometers.
- *Gyroscopes* for measuring angular velocity. Three-axis, two-axis, and one-axis gyros are commonly available. Gyroscopes are relatively immune to environmental interferences and, therefore, have been widely accepted in medical devices [20].
- *Thermal sensors*, a family of sensors that are used to measure temperatures or heat fluxes [19].

- *Electrodes* for monitoring cardiac activity (ECG), brain activity (EEG), respiratory activity (electrical impedance plethysmogram – *EIP*), muscle activity (electromyogram – *EMG*), and emotions (galvanik skin response – *GSR*). They must be applied directly on the skin, typically with disposable adhesive leads that contain a drop of conductive gel.
- *Photoplethysmography (PPG) sensors,* they are used as an indirect method to measure cardiovascular parameters such as pulse rate, blood oxygenation, and blood pressure [21]. They are realized as clips with a light emitting diode (LED) and a photosensible sensor placed at the two terminals. The clip is usually attached to the earlobe or the finger. The operating principle is based on the fact that the blood absorbs or reflects part of the emitted light and the variation of the blood volume caused by heart beats modulates the amount of transmitted or reflected light.

1.11   BSN Application Domains

Comprehensive overviews of several BSN applications can be found in Refs. [22–24]. A few surveys on wearable sensor-based systems have been published to date. For example, in Ref. [22] the focus of the survey is on the functional perspective of the analyzed systems (i.e. what kind of applications they target). In this work, systems are divided into commercial products and research projects, and also grouped on the basis of hardware characteristics: Wired electrode-based, smart textiles, wireless mote-based, and based on sensors found in commercial smartphones. In another frequently cited survey work [23], the attention is focused on the hardware components and the application scenarios. Analyzed projects are classified into (i) in-body (implantable), (ii) on-body medical, and (iii) on-body nonmedical systems.

   Hence, to provide a different point of view, in the following, we will introduce a categorization on the main application domains in which the BSN technology can play a critical role. Moreover, a summary of some literature BSN systems is reported in Table 1.2.

   As aforementioned, BSNs enable a very wide range of application scenarios. We can categorize them into different application domains:

- e-Health
- e-Emergency
- e-Entertainment
- e-Sport
- e-Factory
- e-Sociality

e-Health applications include physical activity recognition, gait analysis, post-trauma rehabilitation after surgeries, cardiac and respiratory diseases prevention

Table 1.2 Summary of representative BSN systems.

Project title	Application domain	Sensors involved	Hardware description	Node platform	Communication protocol	OS/programming language
Real-time Arousal Monitor [25]	Emotion recognition	ECG, respiration, temp., GSR	Chest-belt, skin electrodes, wearable monitor station, USB dongle	Custom	Sensors connected through wires	n/a/C-like
LifeGuard [26]	Medical monitoring in space and extreme environments	ECG, blood pressure, respiration, temp., accelerometer, SpO_2	Custom microcontroller device, commercial biosensors	XPod signal conditioning unit	Bluetooth	n/a
Fitbit® [27]	Physical activity, sleep quality, heart monitoring	Accelerometer, heart rate	Waist/wrist-worn device, PC USB dongle	Fitbit® node	RF proprietary	n/a
VitalSense® [28]	In- and on-body temperature, physical activity, heart monitoring	Temp., ECG, respiration, accelerometer	Custom wearable monitor station, wireless sensors, skin electrodes, ingestible capsule	VitalSense® monitor	RF proprietary	Windows mobile
LiveNet [29]	Parkinson symptom, epilepsy seizure detection	ECG, Blood pressure, respiration, temp., EMG, GSR, SpO_2	PDA, microcontroller board	Custom physiological sensing board	Wires, 2.4 GHz radio, GPRS	Linux (on PDA)

AMON [30]	Cardiac-respiratory diseases	ECG, blood pressure, temp, accelerometer, SpO$_2$	Wrist-worn device	Custom wrist-worn device	Sensors connected through wires –GSM/UMTS	C-like/JAVA (on the server station)
MyHeart [31]	Prevention and detection of cardio vascular diseases	ECG, respiration, accelerometer	PDA, textile sensors, chest-belt	Proprietary monitoring station	Conductive yarns, Bluetooth, GSM	Windows mobile (on the PDA)
Human++ [32]	General health monitoring	ECG, EMG, EEG	Low-power BSN nodes	ASIC	2.4 GHz radio/ UWB modulation	n/a
HealthGear [33]	Sleep apnea detection	Heart rate, SpO$_2$	Custom sensing board, commercial sensors, cell phone	Custom wearable station (includes XPod signal conditioning unit)	Bluetooth	Windows mobile (on the mobile phone)
TeleMuse* [34]	Medical care and research	ECG, EMG, GSR	ZigBee wireless motes	Proprietary	IEEE 802.15.4/ ZigBee	C-like
Polar* Heart Rate Monitor [35]	Fitness and exercise	Heart rate, altimeter	Wireless chest-belt, watch monitor	Proprietary watch monitor	Polar OwnCode* (5 kHz) – coded transmission	n/a

and early detection, remote elderly assistance and monitoring, sleep quality monitoring and sleep apnea detection, and even emotion recognition [36].

e-Emergency refers to applications, e.g. for supporting firefighters and response teams in large-scale disasters due to earthquakes, landslides, and terrorist attacks [37].

e-Entertainment domain refers to human–computer interaction systems typically based on BSNs for real-time motion and gesture recognition, eye tracking, and, more recently, mood and emotion recognition [38, 39].

e-Sport applications are related to the e-Health domain, although they have a nonmedical focus. They include personal e-fitness applications for amateur and professional athletes as well as enterprise systems for professional fitness clubs and sport teams offering advanced performance monitoring services for their athletes [40].

e-Factory is a slowly emerging domain involving industrial process management and monitoring, and workers' safety and collaboration support [41].

Finally, the *e-Sociality* domain involves the recognition of human emotions and cognitive states to enable new forms of social interactions. An interesting example is a system for tracking interactions between two meeting people by detecting, in a collaborative fashion, handshakes and, subsequently, monitoring their social and emotional interactions [42].

1.12 Summary

This chapter has provided an overview of the current state-of-the-art of the BSN domain. We have first introduced the motivations for the BSN technology. We then provided a description of the most important hardware and software requirements of BSN systems, typical m-Health system architecture and, more in detail, the common schematic architecture of a wireless sensor node. In addition, most popular BSN network topologies, communication protocols and standards, and commercial sensor platforms have been introduced. Furthermore, particular attention has been given to the main biophysiological signals and the corresponding physical sensors for their acquisition. Finally, the chapter has provided a categorization of the most relevant BSN application domains and summarized a number of related commercial products and research projects.

References

1 Movassaghi, S., Abolhasan, M., Lipman, J. et al. (2014). Wireless body area networks: a survey. *IEEE Communications Surveys & Tutorials* 16 (3): 1658–1686.

2 Yang, G.Z. ed. (2006). *Body Sensor Networks*. Springer-Verlag.

3 Gravina, R., Alinia, P., Ghasemzadeh, H., and Fortino, G. (2017). Multi-sensor fusion in body sensor networks: state-of-the-art and research challenges. *Information Fusion* 35: 68–80.

4 Kuryloski, P., Giani, A., Giannantonio, R. et al. (2009). DexterNet: an open platform for heterogeneous body sensor networks and its applications. *Proceedings of the Int'l Conference on Body Sensor Networks (BSN 2009)*, Berkeley, CA (3–5 June 2009).

5 International Telecommunication Union (1992). "ARTICLE 1 – Terms and Definitions" – "Industrial, scientific and medical (ISM) applications (of radio frequency energy): operation of equipment or appliances designed to generate and use locally radio frequency energy for industrial, scientific, medical, domestic or similar purposes, excluding applications in the field of telecommunications". http://life.itu.int/radioclub/rr/art1.pdf (accessed 10 June 2017).

6 Venkatesh, C. and Anandamurugan, S. (2010). Increasing the lifetime of wireless sensor networks by using AR (aggregation routing) algorithm. *IJCA Special Issue on MANETs* (4): 180–186.

7 IEEE 802.15.4 Website. http://www.ieee802.org/15/pub/tg4.html (accessed 5 June 2017).

8 ZigBee Website. www.zigbee.org (accessed 5 June 2017).

9 Bluetooth Website. www.bluetooth.com (accessed 10 June 2017).

10 Bluetooth Low Energy Website. https://www.bluetooth.com/what-is-bluetooth-technology/how-it-works/le-p2p (accessed 5 June 2017).

11 ANT Website. www.thisisant.com (accessed 7 June 2017).

12 IEEE 802.15 WPAN Task Group 6 Website. http://www.ieee802.org/15/pub/TG6.html (accessed 8 June 2017).

13 Augimeri, A., Fortino, G., Galzarano, S., and Gravina, R. (2011). Collaborative body sensor networks. *Proceedings of the IEEE International Conference on Systems, Man and Cybertnetics (SMC2011)*, Anchorage, AL (9–12 October 2011).

14 Narayanan, R., Sarath, T., and Vineeth, V. (2016). Survey on motes used in wireless sensor networks: performance & parametric analysis. *Wireless Sensor Network* 8: 51–60.

15 Levis, P., Gay, D., and Culler, D. (2004). Bridging the Gap: Programming Sensor Networks with Application Specific Virtual Machines. *UC Berkeley Tech Rep. UCB//CSD-04-1343*.

16 Mica2 Datasheet. https://www.eol.ucar.edu/isf/facilities/isa/internal/CrossBow/DataSheets/mica2.pdf (accessed 10 October 2016).

17 TelosB Datasheet. http://www.memsic.com/userfiles/files/Datasheets/WSN/telosb_datasheet.pdf (accessed 5 June 2017).

18 Shimmer Platform Website. www.shimmersensing.com (accessed 11 June 2017).

19 Lewis, F.L. (2004). Wireless sensor networks in smart environments: technologies, protocols, applications. In: *Smart Environments: Technologies, Protocols, Applications* (ed. D.J. Cook and S.K. Das). Wiley Blackwell.

20 Madni, A.M., Wan, L.A., and Hammons, S. (1996). A microelectromechanical quartz rotational rate sensor for inertial applications. *Proceedings of the IEEE Aerospatial Applications Conference*, Aspen, CO (3–10 February 1996).

21 Fortino, G. and Giampà, V. (2010). PPG-based methods for non invasive and continuous blood pressure measurement: an overview and development issues in body sensor networks. *2010 IEEE International Workshop on Medical Measurements and Applications, MeMeA 2010 – Proceedings*, Ottawa, ON (30 April to 1 May 2010), Art. No. 5480201, pp. 10–13.

22 Pantelopoulos, A. and Bourbakis Nikolaos, G. (2010). A survey on wearable sensor-based systems for health monitoring and prognosis. *IEEE Transactions on Systems, Man and Cybernetics* 40 (1): 1–12.

23 Ullah, S., Khan, P., Ullah, N. et al. (2009). A review of wireless body area networks for medical applications. *International Journal of Communications, Network and System Sciences* 2 (8): 797–803.

24 Hao, Y. and Foster, R. (2008). Wireless body sensor networks for health-monitoring applications. *Physiological Measurement* 29: 27–56.

25 Grundlehner, B., Brown, L., Penders, J., and Gyselinckx, G. (2009). The design and analysis of a real-time, continuous arousal monitor. *Sixth International Workshop on Wearable and Implantable Body Sensor Networks*, Berkeley, CA (3–5 June 2009), pp. 156–161.

26 Mundt, C.W., Montgomery, K.N., Udoh, U.E. et al. (2005). A multiparameter wearable physiological monitoring system for space and terrestrial applications. *IEEE Transactions on Information Technology in Biomedicine* 9 (3): 382–391.

27 Fitbit Website. www.fitbit.com (accessed 15 June 2017).

28 VitalSense Integrated Physiological Monitor Website. http://www.actigraphy.com/solutions/vitalsense (accessed 8 June 2017).

29 Sung, M., Marci, C., and Pentland, A. (2005). Wearable feedback systems for rehabilitation. *Journal of NeuroEngineering and Rehabilitation* 2: 17.

30 Anliker, U., Ward, J.A., Lukowicz, P. et al. (2004). AMON: a wearable multiparameter medical monitoring and alert system. *IEEE Transactions on Information Technology in Biomedicine* 8 (4): 415–427.

31 Luprano, J., Sola, J., Dasen, S. et al. (2006). Combination of body sensor networks and on-body signal processing algorithms: the practical case of MyHeart project. *Proceedings of the International Workshop Wearable Implantable Body Sensor Networks*, Cambridge, MA (3–5 April 2006), pp. 76–79.

32 Gyselinckx, B., Van Hoof, C., Ryckaert, J. et al. (2005). Human++: autonomous wireless sensors for body area networks. *Proceedings of the IEEE Custom Integrated Circuits Conference*, San Jose, CA (18–21 September 2005), pp. 13–19.

33 Oliver, N. and Flores-Mangas, F. (2006). HealthGear: a real-time wearable system for monitoring and analyzing physiological signals. Microsoft Research. *Tech. Rep. MSR-TR-2005-182*.

34 Biocontrol Systems Website. www.biocontrol.com (accessed 12 June 2017).

35 Polar Electro Website. www.polar.com (accessed 12 June 2017).

36 Gravina, R., Andreoli, A., Salmeri, A. et al. (2010). Enabling multiple BSN applications using the SPINE framework. *Proceedings of the International Conference on Body Sensor Networks, BSN 2010*, Singapore, pp. 228–233 (7–9 June 2010). IEEE Computer Society.

37 Lorincz, K., Malan, D.-J., Fulford-Jones, T. et al. (2004). Sensor networks for emergency response: challenges and opportunities. *IEEE Pervasive Computing* 3 (4): 16–23.

38 Terada, T. and Tanaka, K. (2010). A framework for constructing entertainment contents using flash and wearable sensors. *Proceedings of the 9th International Conference on Entertainment computing, ICEC'10*, Seoul, Korea (8–11 September 2010), pp. 334–341. Springer-Verlag.

39 Gravina, R. and Fortino, G. (2016). Automatic methods for the detection of accelerative cardiac defense response. *IEEE Transactions on Affective Computing* 7 (3): 286–298.

40 Coyle, S., Morris, D., Lau, K. et al. (2009). Textile-based wearable sensors for assisting sports performance. *Proceedings of the International Conference on Body Sensor Networks, BSN 2009*, Berkeley, CA, USA (3-5 June 2009), pp. 228–233. IEEE Computer Society.

41 Huang, J.-Y. and Tsai, C.-H. (2007). A wearable computing environment for the security of a large-scale factory. *Proceedings of the 12th International Conference on Human-Computer Interaction: Interaction Platforms and Techniques, HCI'07*, Beijing, China (22–27 July 2007), pp. 1113–1122. Springer-Verlag.

42 Augimeri, A., Fortino, G., Rege, M. et al. (2010). A cooperative approach for handshake detection based on body sensor networks. *Proceedings of the IEEE International Conference on Systems, Man, and Cybernetics, SMC 2010*, Istanbul, Turkey (10–13 October 2010), pp. 281–288. IEEE Press.

2

BSN Programming Frameworks

2.1 Introduction

Beside the technological hardware developments in terms of system integration, miniaturization, circuitry design, and energy efficiency, developing effective and efficient software applications is the main key factor for wearable systems to emerge and turn from research prototypes into powerful cutting-edge real-world products.

However, building high quality and efficient applications is a hard task to be accomplished without proper programming skills and flexible development tools. This is a very limiting factor, especially in light of the fact that developers of BSN applications may be expert in specific scientific fields (e.g. biology, medicine, and fitness) rather than in networking or embedded programming. As a result, there is an evident need for appropriate methodologies and abstractions capable of improving and simplifying the BSN system development, deployment, and maintenance processes.

This chapter investigates problems and challenges involved in programming BSNs and discusses the importance of adopting high-level programming abstractions and software tools through which developers are able to overcome the difficulties in managing such distributed and resource-constrained embedded environments. Moreover, it provides the state-of-the-art of middleware and programming frameworks by focusing on both capabilities and lack of proper functionalities needed for facing today's and future challenges in BSN application development.

2.2 Developing BSN Applications

Despite more than a decade of research in the BSN field, programming complexity is still one of the challenging issues guilty of hindering a wider diffusion of such systems in the real world.

Wearable Computing: From Modeling to Implementation of Wearable Systems Based on Body Sensor Networks, First Edition. Giancarlo Fortino, Raffaele Gravina, and Stefano Galzarano.

Implementing software on BSN-based systems requires the developers to face many different programming aspects ranging from efficiently managing the very limited hardware resources (power, memory, and computational capability) of the sensor platforms to translating the global distributed in-network application behavior into a per-node set of functions and interacting routines. Dealing with platform-level, network-level, and application-level implementation and debugging steps, without flexible development supporting tools, very likely leads to time-consuming and error-prone tedious tasks prior to having the end-user application ready for deployment.

Unfortunately, a standard and common approach able to effectively fill the gap between the complexity of the routines for managing sensor platforms and network infrastructure, and the high-level requirements of the desired user applications has not been defined yet. Furthermore, with the ever-increasing application complexity due to the more and more advanced functionalities and services provided to the users, the need for integrating different sensor architectures with other types of devices will lead to further challenges in terms of platform interoperability in more heterogeneous and pervasive environments.

Some of the chapters will give a more in-depth discussion of today's scenarios in which the typical single-user BSN system is required to be integrated with other computing paradigms and infrastructures in order to build smarter human-centered environments and enable more complex services for improving the human well-being. To accommodate these new scenarios, the development of such enhanced BSNs will entail the adoption of novel systematic design approaches based on high level and preferable standardized abstractions required to implement, for instance, agent-oriented BSNs (see Chapter 6), multi-BSN collaborative systems (Chapter 7), BSN and building sensor network integration (Chapter 8), and cloud-enabled wearable systems (Chapter 9).

As of today, one of the following development methodologies can be adopted to build BSN applications [1]: (i) application- and platform-specific programming, (ii) automatic code generation, and (iii) middleware-based programming.

2.2.1 Application- and Platform-Specific Programming

Application- and platform-specific programming refers to developing applications that tend to be tailored for a specific purpose. Since they are expressly coded to meet specific requirements and accomplish well-defined tasks, they can be optimized for achieving high performance once deployed. By means of standard programming languages, like C, and by making use of platform-specific Application Programming Interfaces (APIs), developers implement their applications directly on top of a particular operating system or software stack. In such a way, the final result is a single software program consisting of the application logic tightly coupled with the network protocol routines and other services because of direct interactions with the embedded operating system and the hardware controlling components. Although such a design strategy can lead to a highly optimized code in terms of energy consumption and computational performance, the strong coupling between the application and the underlying supporting software is a main issue. This leads to a monolithic piece of code specifically conceived to accomplish a fixed task and usually targeting a single sensor platform, thus resulting in a rigid and poorly reusable infrastructure with no easy-to-reuse software component that could actually be shared by different applications. Although this approach may still be a viable solution for developing quite simple applications, today's complex systems are hard to be implemented without proper versatile development tools. Indeed, the currently available platform APIs tend to leave to the developers many low-level aspects related to the hardware control (e.g. the access to onboard sensor drivers), event handling, as well as in-node job scheduling and code optimization for an efficient use of the scarce node resources. Also, some operating system primitives do not make common BSN functionalities (i.e. sensor configuration and sampling, multinode communication patterns, or distributed data processing) available as ready-to-use and customizable software components. As a result, coding the global application logic into individual node's behavior implies coping with cumbersome tasks like inter-node process synchronization and data integrity and explicitly interfacing with the node-supported network protocols to exchange and parse messages. Therefore, BSN developers have to spend most of their development time in implementing ad-hoc routines dealing with low-level details rather than focusing on the application core logic. Since the implementation is bound to a specific sensor node architecture and a specific set of sensor drivers, the final code is not reusable or easily modifiable in case a different platform is required to be used.

The difficulties and limits in developing platform-specific applications directly on top of a sensor platform's operating system have also been investigated in Ref. [2], which specifically takes into consideration TinyOS [3], MANTIS [4], and the Ember ZigBee stack [5].

Early works on BSNs have focused on small and simple applications with no relevant development issues. However, as already discussed, when application complexity increases, the lack of proper high-level programming tools becomes a strong limiting factor. This is particularly true in light of the fact that many recent application domains are demanding for multiple interconnected Internet-based sensor networks requiring more complex multiplatform applications enabling the claimed paradigm of the Internet of Things (IoT) [6]. In this very near future scenario, broader and more powerful programming interfaces are of paramount importance for better supporting more pervasive computing systems. On the basis of these considerations, there is a strong interest in using software instruments capable of simplifying application development on BSNs.

2.2.2 Automatic Code Generation

The automatic code generation approach aims at solving the problem of making a certain application available for different sensor platforms without tackling multiple manual porting procedures which, depending on the complexity of the application, may be very time-consuming. The technique consists in specifying the application logic through a well-defined platform-independent modeling language, which abstracts away any low-level details related to both hardware and operating system. Subsequently, starting from the defined high-level abstractions, a tailor-made translator tool interprets the application model and generates a source code that can only run on a specific hardware platform and operating system. Thus, such an approach requires that each platform has its own tool for translating the high-level modeling constructs into its low-level programming language application. The most annoying drawback with this approach is the need for recompiling and reflashing the firmware into every single sensor node whenever a change is made in the application model, unless over-the-air (OTA) programming is supported by the platform.

2.2.3 Middleware-Based Programming

Middleware-based programming allows developers to speed-up and ease the application development tasks by benefiting from the use of (i) well-defined high-level abstractions, representing the interface to developers, and (ii) a middleware providing proper runtime mechanisms implementing such abstractions.

Programming frameworks based on middleware support the whole application development (including deployment, execution, and maintenance) by hiding the complexity and the heterogeneity of the sensor platforms, so that the work of developers is facilitated leading to simpler programming, increased code reuse, and easier maintenance. A typical framework solution usually comes with the following components (see Figure 2.1):

1) *Programming abstraction*: it provides a programming interface to a specific development paradigm and built-in functionalities for an easier management of physical and basic-software resources (such as storage, sensing, communication, and operating system). Since the final application is defined in terms of well-defined high-level constructs representing the interface to the BSN functionalities, the developer can focus on the application logic rather than dealing with the implementation of lower level mechanisms.

2) *Middleware services and functions*: a set of reusable routines in charge of providing the actual implementation of the high-level constructs constituting the programming abstraction. They include the middleware common core functionalities and networking mechanisms to perform the user-defined application execution.

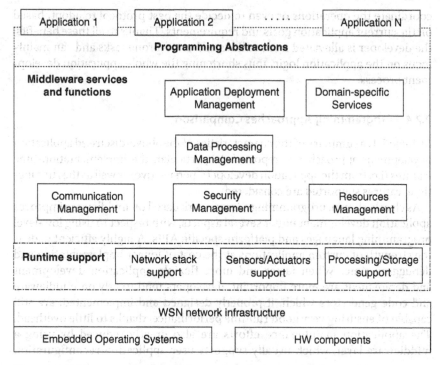

Figure 2.1 Reference model of a middleware-based programming framework.

3) *Runtime support*: serves as a specific execution environment for supporting the services and functions. In practice, it performs the interaction between the middleware layer and the sensor platforms, i.e. the embedded operating systems and the hardware components.

The middleware-based development approach is gaining more and more attention in the BSN domain and is currently considered as the most effective one in bridging the software gap between the complexity of the routines for managing the hardware/operating system/network stack layers of the sensor platforms and the requirements of the application logic. Therefore, a middleware is generally designed as a distributed software layer running on each sensor node and in charge of providing a set of interfaces and services to the upper layers in order to hide low-level details of the underlying system architecture and the related networking protocols. In particular, it is responsible for the actual execution of the user-defined application by "translating" the high-level programming abstractions into real running functions aiming at, for instance, extracting, collecting, processing, and transporting data within and across nodes. At the same time, it may handle some low-level management routines for constantly controlling platform resources and network status to better coordinate the operations or even to decide the best protocol to adopt, based on the current application goals and requirements. Thanks to all these benefits, the developer is alleviated from tedious and error-prone tasks and can mainly focus on the application logic, thus shortening the whole application development process.

2.2.4 Programming Approaches Comparison

In Table 2.1, a summary of the characteristics of the above-discussed application development approaches is reported. In particular, the implementation-time features (i.e. from the application developers' perspective) as well as the running-time features supported are considered.

As clearly shown, programming a framework based on middleware improves application development under several aspects, with respect to using low-level programming languages and platform-specific APIs. A highly efficient code is the strength of custom applications at the cost of longer implementation and debugging time. When faster and more flexible application development and deployment are more important, developers tend to rely on middleware and code generators which, if properly designed and implemented, are still capable of ensuring very good runtime performance, thanks to little overhead. The application maintenance efforts are also greatly reduced by using a middleware layer, which usually supports user application reconfiguration

Table 2.1 BSN-application development approaches comparison.

	Application-specific and platform-specific programming	Automatic code generation	Middleware-based programming
High-level application modeling		✓	✓
Rapid prototyping		✓	✓
Ease of debugging		✓	✓
Quick application development		✓	✓
Application reconfiguration at runtime			✓
Code efficiency	✓	✓	✓
System interoperability			✓
Software reusability		✓	✓

without the need for reflashing an updated firmware into each single node. This is accomplished by means of proper messages interpreted by the middleware running on the nodes, so as to prevent developers to physically access the devices. On the contrary, both low-level programming and code generator-based approaches do not provide such a feature, since they generate new firmware, which need to be manually uploaded on each node, unless a sensor platform providing an OTA programming functionality is employed. Another important requirement in the BSN is system interoperability that is the property of different applications to cooperate across heterogeneous platforms. When developed in a middleware environment, the common messaging protocol at the high level offers the best support to this purpose, whereas in the other approaches, developers have to put much more efforts and time in order to achieve similar results. Finally, the design strategy of building systems targeting specific applications generates rigid software architecture with no reuse of software components or infrastructure.

2.3 Programming Abstractions

As already mentioned, the programming abstractions provide the primary interface for the developers and represent the basis for the programming paradigm supported by the middleware running on the sensor network infrastructure. These mechanisms can include high-level constructs for defining several operations like sensing, sensor reading aggregation, and data-flow control, computation, and communication. If properly conceived, such abstractions greatly relieve application developers from directly dealing with tedious low-level details such as resource management, network protocols, and power management, among others. For addressing sensor network programming issues and supporting developers in a fast and effective application development, in the last decade, many frameworks for sensor networks have been proposed, focusing on different application aspects. Depending on the specific scope of applicability, each of them provides a well-defined programming paradigm along with its related high-level abstractions. To some extent, most of these high-level approaches can also be employed for building BSN applications. However, as it will be discussed later in this chapter, a BSN system poses different challenges and demands more specific requirements to be fulfilled. Thus, appropriate programming paradigms and supporting tools specifically designed to accommodate such peculiar needs are required to better exploit the potentialities of BSNs.

In the following, a list of existing programming paradigms, and related supporting frameworks, for sensor networks is reported.

Task-oriented paradigm (SPINE2 [1, 7], Titan [8], and ATaG [9]): the task-oriented approach aims at providing an easy and effective way for developing distributed applications as a composition of basic functional blocks, tasks. Each task usually performs a well-specified operation such as a data-processing function or a sensor sampling. By means of such a data-flow-oriented chain of interconnected tasks (data flow from sensors to processing results), developers are able to quickly translate the application logic into a high-level, modular, and easily reconfigurable representation, which is then automatically executed over the sensor network by means of a proper runtime system, provided as a common middleware layer running on every node. This intuitive programming model is particularly suitable for distributed signal processing, which represents the main application in the BSN context.

Agent-based paradigm (MAPS [10–12], AFME [13], Agilla [14], SensorWare [15], and actorNet [16]): the agent-based programming model is associated with the notion of multiple, desirable lightweight, agents migrating from node to node performing part of a given task and collaborating each other to implement a global distributed application. An agent could read sensor values, actuate devices, and send radio packets. The users do not have to define a per-node logic, but an arbitrary number of agents and their behavior, specifying how they collaborate to accomplish the needed tasks on the network. According to this model, the programming paradigm provides users with high-level constructs to define agents' characteristics by hiding how communication and mobility are actually implemented. Such a paradigm allows developers to build distributed, modular applications that can be easily reconfigured and relocated by means of a mobile code.

Function-based paradigm (SPINE [17], C-SPINE [18, 19], RehabSPOT [20], and CodeBlue [21]): Not based on a specific formalism for abstracting data or task, these frameworks provide developers with customizable functions as main programming interfaces for data collection, processing, and displaying. They come with easily reusable libraries and tools conceived to specifically address and standardize the core challenges of sensor-based system design within a particular application domain. Moreover, since there is no complex execution engine for "translating" high-level abstractions, a very lightweight and flexible middleware guarantees high runtime performance.

Macroprogramming paradigm (ATaG [9], Logical Neighborhoods [22], Kairos [23], and Regiment [24]): this approach is for developing highly distributed applications since it easily allows the definition of the global behavior of the whole sensor network, rather than single actions related to individual nodes. This approach has been conceived for dealing with WSNs constituted by a large number of nodes, such that the complexity in coordinating their actions makes applications quite difficult to be designed in an effective way. The same effectiveness cannot be considered when applied to the BSNs. Macroprogramming generally has some language constructs for abstracting embedded system's details, communication protocols, node collaboration, and resource allocation. Moreover, it provides mechanisms through which sensors can be divided into logical groups on the basis of their locations, functionalities, or roles. Then, the programming task decreases in complexity because programmers have only to specify what kind of collaborations exist between groups, whereas the underlying execution environment is in charge of translating these high-level conceptual descriptions into actual node-level actions. Thanks to these high-level concepts, any domain expert not skilled in programming can develop their own application by simply defining the whole system behavior through concepts and terms they are familiar with.

Model-based paradigm [2]: it allows developers to define proper models representing the desired behavior of an application. Usually, such an approach consists in making use of a well-defined modeling language (such as finite state machines and flow charts) and a tool capable of generating a low-level code for a specific target platform starting from the model. Although it represents a standard methodology for several domains, such as automotive electronics, its employment in the context of WSN/BSN has not been widely investigated yet.

Application-driven paradigm (MiLAN [25]): middlewares belonging to this model aim to provide services to applications according to their needs and requirements, especially for QoS and reliability of the collected data. They allow programmers to directly access the communication protocol stack for adjusting the network functions to support and satisfy the requested requirements.

Database paradigm (TinyDB [26], Cougar [27], and SINA [28]): The database model lets users view the whole sensor network as a virtual relational distributed database system allowing a simple and easy communication scheme between users and network. Through the adoption of easy-to-use languages, the users have the ability to make intuitive queries for extracting the data of interest from the sensors. The most common way for querying networks is making use of a SQL-like language, a simple semi-declarative style language. This paradigm is mainly designed to collect data streams, with the limitation of providing only approximate results. Moreover, it is not suitable to support real-time applications (usually a must in BSNs) because it lacks a time–space relationship between events.

Virtual machine paradigm (Maté [29], DAViM [30], and DVM [31]): Virtual machines (VMs) have been generally adopted for software emulating a guest system running on top of a real host. In the WSN context, VMs are used for allowing a broad range of applications to run on different platforms without worrying about the underlying architecture characteristics. User applications are coded with a simple set of instructions that are interpreted by the VM execution environment. Unfortunately, this approach suffers from the performance overhead that the instructions' interpretation introduces.

2.4 Requirements for BSN Frameworks

BSN applications, despite their diversification, share several common tasks on top of which the application-specific logic is implemented. A correct and clear identification of such tasks is essential to realize an effective and usable BSN programming framework.

Table 2.2 summarizes the results of an in-depth analysis of research projects and technological prototypes to identify the very essential set of tasks commonly needed by BSN applications.

The tasks reported in Table 2.2 should be provided by a framework for the development of BSN applications, for instance by means of programming abstractions and tools. In addition, such a framework should be designed to meet specific (functional and nonfunctional) requirements in terms of effectiveness, efficiency, and usability to be actually capable of facilitating the development of well-structured and resource-efficient applications with less effort in terms of development time and application programming complexity. The resulting source code should be more reusable, easier to maintain, and supported by tools for application management. Supporting heterogeneous sensor platforms is also relevant; hence, system interoperability is a desirable requirement, too. Finally, privacy and security are highly important requirements because it is important to protect identifiable and sensitive data such as the ones coming from physiological, possibly medical-relevant signals. In Table 2.3 we have reported the aforementioned requirements that we deem fundamental for a BSN-specific software framework.

Programming Effectiveness is the ability of the framework to provide effective and specific support for application programming, debugging, and testing. In practice, it is realized by programming abstractions, software engineering methods, and debugging and testing tools. More specifically:

- *Programming abstractions* help developers to focus on core application aspects by providing higher level functionalities, as already discussed. In the domain of BSN development, it is particularly relevant to find (i) tunable sensor drivers (to adjust, possibly at runtime, sampling rate, sensitivity, and range, or to enable/disable only certain channels of a multichannel sensor), (ii) flexible data structures (to handle different data types), (iii) flexible communication APIs (different applications typically require different packet lengths and structures in terms of data payload), and (iv) parameterized processing functions (to set functions' parameters without hard-coding their values).
- *Software engineering* methods use component-based (object-like) approaches to support rapid BSN application prototyping. A software framework should provide predefined (ready to use) BSN-specific components that are common to most applications; this will help developers to reach prototypes in shorter time. Examples of such common components

Table 2.2 Common tasks of BSN applications.

Task	Description
Sensor sampling	Sensor sampling is typically the first step of BSN application development. Each application has different requirements and each physiological signal has its own characteristics, so it is strategic to properly tune the sensor sampling rate, as it eventually influences the amount of raw data generated and the quality of the extracted information.
In-node data processing	Pattern recognition and data mining algorithms often need preprocessing of raw data to increase its quality and reduce its amount. Raw signals are typically filtered (e.g. to mitigate the effect of noise sources) and features are extracted in the processing workflow before inferring higher level information. In-node and real-time feature extraction is an important task to reduce wireless traffic and computation workload on the coordinator.
Runtime sensor configuration	Configuring at runtime each sensor node is useful because application demand can change during its execution, so as to allow for dynamic application behavior. For example, under certain circumstances, it might be convenient to reduce the sampling rate of a specific sensor, or even disable its data transmission.
Node synchronization	Many distributed signal-processing algorithms require multiple nodes to be sampled synchronously (i.e. at the same actual time intervals), to ensure consistency of data observation and underlying events. Nodes clocks in these cases must be kept synchronized to preserve synchronized sampling of individual sensor signals.
Duty-cycling	Duty-cycling is a mechanism for controlling the activation of hardware resources (typically radio, sensor transducers, and microcontroller) only when actually needed, to reduce power consumption and hence increase battery lifetime of the sensor node.
Application-level communication protocol	As the application complexity increases, interactions among sensor nodes and between sensor nodes and the coordinator become diversified. For instance, communication involves sensor node discovery/advertisement, requests for sensing and processing activation and configuration, raw and processed sensor data transmission, and event delivery. In this scenario, a flexible application-level communication protocol would better support the application development.
High-level processing	BSN application services often require pattern recognition and classification algorithms to enable fine interpretation of BSN-generated asynchronous events and periodic data to extract meaningful information and mine high-level knowledge.

Table 2.3 Requirements for BSN frameworks.

Requirement	High-level techniques
Programming effectiveness	Programming abstractions, software engineering methods, debugging and testing tools
System efficiency	Resource management optimization
System interoperability	Application-level communication protocol and adapters for heterogeneous platform support
System usability	User-friendly BSN management, PC and mobile device-based coordinator
Privacy support	Data encryption and authentication

are signal filters (e.g. FIR filters) to clean or amplify a signal, feature extractors (e.g. average, variance, zero crossing, and signal slope) to reduce the amount of transmitted data, classification algorithms (e.g. k-NN, decision trees) useful as decision support tools, and an application-level communication protocol (e.g. for nodes/services discovery, failure notification, and user data transmission).

- *Debugging and Testing tools* are necessary to verify functional correctness of the application under development. Debugger tools help in locating the causes of known erroneous application behaviors, while testing tools help in verifying the correctness of software components. They may be included with the development environment and can consist of simulators or step-by-step debuggers.

System Efficiency indicates qualitatively the performance of the system in terms of energy, storage, and computational resource management. Built-in tunable power management schemes let adjusting the trade-off between performance, reliability, and system lifetime. Power management aims at improving BSN lifetime, often by means of radio duty-cycling, sensor down-sampling, or by disabling wireless data transmission in favor of local storage.

System Interoperability is the ability of enabling collaboration (i.e. communication, distributed sensing, and processing) among different devices in terms of hardware/software technologies. To exemplify, interoperability scenarios include (i) network formation and communication among devices based on different hardware architecture but programmed using the same language, (ii) interoperability among homogeneous BSN coordinators, and (iii) the ultimate ability of a system to interoperate with fully heterogeneous devices (e.g. Internet through sockets or XML RPC). In practice, it can be achieved with an application-level communication protocol and communication adapters for supporting heterogeneous sensor and coordinator devices.

System Usability is a (nonfunctional) property referring to systems that are easy-to-use for designers, developers, and end users. It is often supported by graphical or API-based BSN management tools running on a remote coordinator (a PC or a mobile device).

Privacy Support is the ability of a system to protect user's confidential information. Encryption and authentication functionalities allow the system to keep such information secret and to ensure access only to authorized parties. Privacy protection is a necessary requirement in every real-world e-Health applications and it can be effectively achieved only when all the system tiers use privacy policies.

As for the programming *abstractions*, on the basis of what was discussed in Section 2.3, it emerges that none of them can be considered as the predominant one. Depending on specific tasks and/or contexts, a certain solution may result as a better choice than others. Most of them have peculiar features specifically conceived for particular application contexts but lack in characteristics useful for more general-purpose uses. For instance, frameworks based on a database approach provide high-level services for data aggregation and querying but not for defining a more general-purpose computation. Hence, the data-centric model is not suitable in domains requiring more sophisticated collaborative sensor data processing over the network. In the specific context of BSN-based systems, most of these frameworks do not allow a distributed data flow management and processing over the network. Fast application reconfiguration and platform independence are two fundamental requirements to be fulfilled by a BSN programming paradigm. Reprogramming a network is a desirable feature for supporting rapid and efficient changes of sensor node behavior. Systems like Deluge [32] and TinyCubus [33] provide code updates by directly loading them over the radio. However, they require the use of a homogeneous hardware/software platform; also, the code transmission is a time- and energy-consuming operation. VMs represent a typical approach for achieving a platform-independent behavior. They allow the development of applications by means of proper instructions, which are interpreted by the VM running on sensor nodes. Unfortunately, this approach requires high computational and memory resources and suffers of poor performance due to the overhead for interpreting the instructions. Moreover, coding an application with the provided instructions is not fast and intuitive (e.g. Maté provides more than a hundred instructions), especially if the application needs frequent changes.

2.5 BSN Programming Frameworks

In the following, a brief description of the main current frameworks and architectures for developing BSN-based systems is presented.

2.5.1 Titan

Titan (Tiny task network) [8] is a programming framework conceived to specifically enable dynamic context recognition on the BSN. A Titan application is represented by a task graph that is defined as a set of interconnected basic blocks, tasks, which are executed over the sensor network by the framework runtime system. In particular, once the whole application is defined, the task network is split into a set of task subnetworks, each of which is assigned and executed on a single node. In case of two tasks placed on different nodes, the data transfer takes place through messages exchanged via an ad-hoc communication protocol. Each task is mapped and executed only on a specific node, unless it will become unavailable during execution, e.g. due to battery depletion. In such a case, the Titan coordinator automatically performs a reallocation of the task by picking one of the remaining running nodes that has sufficient resources to handle that specific task. The middleware is also in charge of accordingly readdressing the inter-task communication based on the previously defined task graph. Titan provides developers with a library of predefined tasks, each representing a specific operation such as a sensor reading, a processing function, or a classification algorithm.

2.5.2 CodeBlue

CodeBlue [21] is a sensor network infrastructure specifically conceived to support medical scenarios ranging from indoor monitoring of patients in medical centers to outdoor disaster emergency management. The final aim is to effectively support highly critical decision support systems by continuously feeding patient information coming from a set of wearable medical sensors (based on TelosB [34] and MicaZ [35]). The middleware platform, built atop TinyOS, is designed to provide high-level services, such as ad-hoc routing, naming, discovery, and security, and is capable of scaling across a wide range of network densities, from sparse clinic environments to mass casualty sites. Mainly focusing on communication services, CodeBlue is based on a flexible publish/subscribe data delivery model in order to provide a common scalable and robust (in case of the temporary loss of radio connectivity) information plane for coordinating medical devices. In particular, sensors publish important data to given channels and coordinator devices (hand-handled or laptop) subscribe to channels of interest.

2.5.3 RehabSPOT

RehabSPOT [20] is a BSN platform based on Sun SPOT sensor nodes [36] designed for facilitating physical therapists' work and improving patients' limb rehabilitation treatment. Based on a three-tier customizable platform, it features adaptive data collection, online processing, and display. In particular, the wearable nodes are organized as a standalone mesh network (first tier) and each of them runs a client software. A coordinator (second tier, usually a PC) is in charge of managing the nodes by forming a star-topology network and performing real-time display and online processing. Finally, an Internet infrastructure (third tier) is designed to upload data from the coordinator to remote servers for off-line analysis.

2.5.4 SPINE

SPINE [17, 37] is an open-source BSN framework for effective development of distributed signal processing. It provides a variety of built-in sensor drivers, signal-processing functions, and flexible data communication protocols. Also, its architecture allows for easy integration of new customized sensor drivers and processing functionalities. SPINE currently supports the most popular programmable sensor node platforms running TinyOS, i.e. Tmote Sky/TelosB, MicaZ, and Shimmer [38]. In addition, there exist SPINE implementations for (i) ZigBee devices based on the TI Z-Stack and (ii) the Java Sun SPOT sensors [36]. A more in-depth description of SPINE is presented in Chapter 3.

2.5.5 SPINE2

SPINE2 [1, 7], evolved from SPINE, is a platform-independent framework designed around a task-oriented high-level programming approach. According to this paradigm, a signal-processing application is defined in terms of a network of tasks, where each task (available from a library of tasks) represents a specific activity, like a sensing operation, a processing function, or a data transfer. Designing applications with a set of basic building blocks enables a more rapid system development, runtime re-configuration, and easier software maintenance. The software architecture of SPINE2, designed by following a software layering approach, is composed of several platform-independent components and a set of platform-dependent modules to access the specific platform resources and services. This leads to an easier and faster porting of SPINE2 to new C-like sensor platforms. A more in-depth description of SPINE2 is presented in Chapter 4.

2.5.6 C-SPINE

C-SPINE [18, 19] is a SPINE-based programming framework specifically designed to support the development of distributed applications over Collaborative BSNs (CBSNs). The C-SPINE architecture includes the SPINE sensor-side and the SPINE base station-side software components, with the addition of specific CBSN architectural components enabling several services providing Inter-CBSN

Communication, BSN Proximity Detection, BSN Service Discovery, BSN Service Selection, and Application-specific Protocols and Services, which specifically support collaborative computing and multisensor data fusion among BSNs. C-SPINE is described in Chapter 7.

2.5.7 MAPS

MAPS [10–12] is a Java-based programming framework enabling agent-oriented programming over sensor networks. It has been widely used for developing a BSN-specific system showing the versatility of such a programming approach. MAPS provides developers a set of fundamental services for programming agents including message transmission, agent creation, agent cloning, agent migration, timer handling, and easy access to the sensor node resources, whereas the agents' behavior is modeled as a multiplane state machine. MAPS is presented in Chapter 6 along with a more general discussion about the benefits of agent-oriented programming approaches for developing BSN systems.

2.5.8 DexterNet

DexterNet [39] is an open-source platform for BSN supporting scalable, real-time human monitoring in indoor and outdoor environments over heterogeneous wearable sensors. The software platform is designed as a three-tier architecture, which includes the following: (i) the body sensor layer (BSL), (ii) the personal network layer (PNL), and (iii) the global network layer (GNL). The first two layers are implemented by using the SPINE framework libraries for managing a single BSN, whereas the third one allows a multiple-PNL communication over the Internet and supports higher level applications for remote data logging and analysis.

2.6 Summary

This chapter discussed the programming issues in sensor networks, with particular regard to the methodologies for efficiently and effectively building applications on BSNs. We have first introduced and compared the different development approaches. We then focused on the most common programming abstractions provided in the literature by highlighting their main peculiarities and features and their applicability in the BSN domain. Furthermore, the requirements for designing effective BSN-specific frameworks have been discussed. Finally, the current available frameworks for developing BSN applications have been briefly described.

References

1 Galzarano, S., Giannantonio, R., Liotta, A., and Fortino, G. (2016). A task-oriented framework for networked wearable computing. *IEEE Transactions on Automation Science and Engineering* 13 (2): 621–638. doi: 10.1109/TASE.2014.2365880.

2 Mozumdar, M.M.R., Lavagno, L., Vanzago, L., and Sangiovanni-Vincentelli, A.L. (2010). HILAC: A framework for hardware in the loop simulation and multi-platform automatic code generation of WSN applications. *2010 International Symposium on Industrial Embedded Systems (SIES)*, Trento Italy (7–9 July), pp. 88–97.

3 Levis, P., Madden, S., Polastre, J. et al. (2005). TinyOS: an operating system for sensor networks. *Ambient Intelligence*, (ed. W. Weber, J.M. Rabaey, and E. Aarts), 115–148. Berlin/Heidelberg: Springer.

4 Bhatti, S., Carlson, J., Dai, H. et al. (2005). MANTIS OS: an embedded multithreaded operating system for wireless micro sensor platforms. *Mobile Network Application* 10 (4): 563–579.

5 EmberZ StackWebsite. http://www.silabs.com/products/development-tools/software/emberznet-pro-zigbee-protocol-stack-software (accessed 6 June 2017).

6 Kortuem, G., Kawsar, F., Fitton, D., and Sundramoorthy, V. (2010). Smart objects as building blocks for the Internet of things. *IEEE Internet Computing* 14 (1): 44–51.

7 Raveendranathan, N., Galzarano, S., Loseu, V. et al. (2012). From modeling to implementation of virtual sensors in body sensor networks. *IEEE Sensors Journal* 12 (3): 583–593.

8 Lombriser, C., Roggen, D., Stager, M., and Troster, G. (2007). Titan: a tiny task network for dynamically reconfigurable heterogeneous sensor networks. In *Kommunikation in Verteilten Systemen (KiVS)*, 127–138. New York: Springer.

9 Bakshi, A., Prasanna, V.K., Reich, J., and Larner, D. (2005). The abstract task graph: a methodology for architecture-independent programming of networked sensor systems. *Proceedings of the 2005 Workshop on End-to-End, Sense-and-Respond Systems, Applications and Services*, Seattle, WA (5 June 2005), pp. 19–24.

10 Aiello, F., Fortino, G., Gravina, R., and Guerrieri, A. (2011). A Java-based agent platform for programming wireless sensor networks. *The Computer Journal* 54 (3): 439–454.

11 Aiello, F., Bellifemine, F., Fortino, G. et al. (2011). An agent-based signal processing in-node environment for real-time human activity monitoring based on wireless body sensor networks. *Journal of Engineering Applications of Artificial Intelligence* 24: 1147–1161.

12 Aiello, F., Fortino, G., Gravina, R., and Guerrieri, A. (2009). MAPS: a mobile agent platform for Java Sun SPOTs. *Proceedings of the 3rd International Workshop on Agent Technology for Sensor Networks (ATSN-09)*, jointly held with the *8th International Joint Conference on Autonomous Agents and Multiagent Systems (AAMAS-09)*, Budapest, Hungary (12 May 2009).

13 Muldoon, C., O'Hare, G.M.P., Collier, R.W., and O'Grady, M.J. (2006). Agent factory micro edition: a framework for ambient applications. *Proceedings of Intelligent Agents in Computing Systems*, ser. Lecture Notes in Computer Science, vol. 3993 (28–31 May 2006), pp. 727–734. Reading: Springer.

14 Fok, C.-L., Roman, G.-C., and Lu, C. (2009). Agilla: a mobile agent middleware for self-adaptive wireless sensor networks. *ACM Transactions on Autonomous and Adaptive Systems* 4 (3): 16:1–16:26.

15 Boulis, A., Han, C.-C., and Srivastava, M.B. (2003). Design and implementation of a framework for efficient and programmable sensor networks. *Proceedings of the 1st International Conference on Mobile Systems, Applications and Services*, San Francisco, CA (5–8 May 2003), pp. 187–200.

16 Kwon, Y., Sundresh, S., Mechitov, K., and Agha, G. (2006). ActorNet: an actor platform for wireless sensor networks. *Proceedings of the 5th International Joint Conference on Autonomous Agents and Multiagent Systems (AAMAS)*, Hakodate, Japan (8–12 May 2006), pp. 1297–1300.

17 Fortino, G., Giannantonio, R., Gravina, R. et al. (2013). Enabling effective programming and flexible management of efficient body sensor network applications. *IEEE Transactions on Human-Machine Systems* 43 (1): 115–133.

18 Fortino, G., Galzarano, S., Gravina, R., and Li, W. (2014). A framework for collaborative computing and multi-sensor data fusion in body sensor networks. *Information Fusion* 22: 50–70.

19 Augimeri, A., Fortino, G., Galzarano, S., and Gravina, R. (2011). Collaborative body sensor networks. *Proceedings of the 2011 IEEE International Conference on Systems, Man, and Cybernetics (SMC)*, Anchorage, AL (9–12 October 2011), pp. 3427–3432.

20 Zhang, M. and Sawchuk, A. (2009). A customizable framework of body area sensor network for rehabilitation. *Second International Symposium on Applied Sciences in Biomedical and Communication Technologies (ISABEL)* (24–27 November 2009), pp. 1–6.

21 Malan, D., Fulford-Jones, T., Welsh, M., and Moulton, S. (2004). Codeblue: an ad hoc sensor network infrastructure for emergency medical care. *Proceedings of the International Workshop on Wearable and Implantable Body Sensor Networks*, London, UK (6 and 7 April 2004).

22 Mottola, L. and Picco, G.P. (2006). Logical neighborhoods: a programming abstraction for wireless sensor networks. In: *Distributed Computing in Sensor Systems* (ed. P.B. Gibbons, T. Abdelzaher, J. Aspnes, and R. Rao), 150–168. Berlin/Heidelberg: Springer.

23 Gummadi, R., Kothari, N., Govindan, R., and Millstein, T. (2005). Kairos: a macro-programming system for wireless sensor networks. *Proceedings of the twentieth ACM symposium on Operating Systems Principles*, Brighton, UK (23–26 October 2005), pp. 1–2.

24 Newton, R., Morrisett, G., and Welsh, M. (2007). The regiment macroprogramming system. *Proceedings of the 6th International Conference on Information Processing in Sensor Networks*, Cambridge, MA (25–27 April 2007), pp. 489–498.

25 Heinzelman, W.B., Murphy, A.L., Carvalho, H.S., and Perillo, M.A. (2004). Middleware to support sensor network applications. *IEEE Network* 18 (1): 6–14.

26 Madden, S.R., Franklin, M.J., Hellerstein, J.M., and Hong, W. (2005). TinyDB: an acquisitional query processing system for sensor networks. *ACM Transactions on Database Systems* 30 (1): 122–173.

27 Bonnet, P., Gehrke, J., and Seshadri, P. (2000). Querying the physical world. *IEEE Personal Communications* 7 (5): 10–15.

28 Srisathapornphat, C., Jaikaeo, C., and Shen, C.-C. (2000). Sensor information networking architecture. *Proceedings 2000. International Workshops on Parallel Processing*, Tokio, Japan (14 September 2000), pp. 23–30.

29 Levis, P. and Culler, D. (2002). Maté: a tiny virtual machine for sensor networks. *SIGOPS Operating Systems Review* 36 (5): 85–95.

30 Michiels, S., Horré, W., Joosen, W., and Verbaeten, P. (2006). DAViM: a dynamically adaptable virtual machine for sensor networks. *Proceedings of the International Workshop on Middleware for Sensor Networks*, New York, pp. 7–12.

31 Balani, R., Han, C.-C., Rengaswamy, R.K. et al. (2006). Multi-level software reconfiguration for sensor networks. *Proceedings of the 6th ACM & IEEE International conference on Embedded Software*, Seoul, Republic of Korea (22–27 October 2006), pp. 112–121.

32 Hui, J.W. and Culler, D. (2004). The dynamic behavior of a data dissemination protocol for network programming at scale. *Proceedings of the 2nd International Conference on Embedded Networked Sensor Systems*, Baltimore, MD (3–5 November 2004), pp. 81–94.

33 Marron, P.J., Lachenmann, A., Minder, D. et al. (2005). TinyCubus: a flexible and adaptive framework sensor networks. *Proceeedings of the Second European Workshop on Wireless Sensor Networks, 2005*, Istanbul, Turkey (31 January–2 February 2005), pp. 278–289.

34 TelosB Datasheet. http://www.memsic.com/userfiles/files/Datasheets/WSN/telosb_datasheet.pdf (accessed 11 June 2017).

35 Mica2 Datasheet. https://www.eol.ucar.edu/isf/facilities/isa/internal/CrossBow/DataSheets/mica2.pdf (accessed 5 June 2017).

36 Sun SPOT Website. www.sunspotdev.org (accessed 13 June 2017).

37 Bellifemine, F., Fortino, G., Giannantonio, R. et al. (2011). SPINE: a domain-specific framework for rapid prototyping of WBSN applications. *Software: Practice and Experience* 41 (3): 237–265. doi: 10.1002/spe.998.

38 Shimmer Website. www.shimmersensing.com (accessed 14 June 2017)

39 Kuryloski, P., Giani, A., Giannantonio, R. et al. (2009). DexterNet: an open platform for heterogeneous body sensor networks and its applications. *Sixth International Workshop on Wearable and Implantable Body Sensor Networks, 2009. BSN 2009*, Berkeley, CA (3–5 June 2009), pp. 92–97.

3

Signal Processing In-Node Environment

3.1 Introduction

The analysis of the state-of-the-art on the BSN domain has highlighted that the development of BSN applications is to date a complex task also due to the lack of programming frameworks with dedicated support to the distinctive requirements of BSN systems.

To support the programming of optimized BSN applications while minimizing the development time and effort, we have designed and realized SPINE (Signal Processing In-Node Environment) [1–3], an open-source domain-specific programming framework for BSNs.

SPINE aims at boosting the prototyping of BSN applications. SPINE enables efficient implementations of signal-processing algorithms for analysis and classification of sensor data through libraries of processing functionalities. It is organized into two interacting macro-components, which are, respectively, implemented on commercially available programmable sensor devices and on the personal coordinator (Android smartphones and tablets, or a personal computer). Communication among these devices is wireless, using Bluetooth or IEEE 802.15.4 standards. The high-level SPINE API (at the coordinator level) allows for dynamic and flexible configuration of sensing and processing functionalities available at the sensor node level. Many biophysical sensors and signal-processing tasks are natively implemented and available to application developers. In addition, the SPINE framework has been carefully designed to allow for very easy integration of new, custom-defined sensor drivers and processing tasks. A key advantage of adopting SPINE is its ability to configure the BSN system based on specific sensing and processing requirements; in this way, the same sensors can be used by different applications without requiring off-line reprogramming before switching from an application to another.

Wearable Computing: From Modeling to Implementation of Wearable Systems Based on Body Sensor Networks, First Edition. Giancarlo Fortino, Raffaele Gravina, and Stefano Galzarano.
© 2018 John Wiley & Sons, Inc. Published 2018 by John Wiley & Son, Inc.

3.2 Background

TinyOS [4] is an event-driven operating system, which provides a programming environment for embedded systems. It has a component-based execution model implemented in the nesC language [5] with a very low memory footprint.

TinyOS concurrency model is based on commands, asynchronous events, deferred computation called tasks, and split-phase interfaces. The function invocation (as command) and its completion (as event) are separated into two phases in interfaces provided by TinyOS. Application user has to write the handler, which is invoked upon the triggering of an event. Commands and event handlers may post a task, which is executed by the TinyOS FIFO scheduler. These tasks are non-preemptive among each other and, thus, run to completion. Only an (asynchronous) event can preempt running tasks. Data race conflicts that arise due to preemption can be solved using atomic sections.

Radio communication in TinyOS follows the Active Messages [6] model, in which each packet on the network specifies the ID of the handler that will be invoked on the recipient nodes. The handler ID is an integer that is carried in the header of the message. When a message is received, the event associated with the handler ID is signaled. Different sensor nodes can associate different receive events with the same handler ID.

3.3 Motivations and Challenges

The development of SPINE, as a domain-specific BSN middleware (MW), is motivated by the need of providing more effective solutions than naïve application-specific programming and a more efficient approach than general-purpose programming frameworks. It has been demonstrated that in the BSN domain, domain-specific frameworks contribute to reduce the development cycle and maintenance since they provide high-level abstractions of network protocols and hardware details, allowing the programmer to focus mainly on the application logic without the burden of carrying the overhead of general-purpose functionalities that are, in practice, not used in the BSN domain (e.g. multi-hop support).

The main challenge during the design of SPINE was to find the most effective trade-off between high-level API definition (i.e. fulfillment of requirements in the BSN programming domain) and the limitations given by strongly resource-constrained sensing devices.

3.4 The SPINE Framework

SPINE is a full-fledged and extensible solution that allows rapid prototyping of BSN-based applications and systems. It allows quick implementation of distributed signal-processing intensive applications by supporting several physiological sensors, in-node and on-coordinator signal-processing utilities, wireless transmission of biosignals, and built-in optimized network and resource management. SPINE is designed as a modular structure to simplify the integration of additional sensor drivers and signal-processing modules; in addition, the framework itself can be tailored and customized by a simple mechanism to combine all the sensing and processing modules altogether, according to specific application requirements. A key advantage of adopting SPINE is its ability to configure the BSN system based on specific sensing and processing requirements; in this way the same sensors can be used by different applications without requiring off-line reprogramming before switching from an application to another. SPINE supports BSN networks that are conceptually organized in a star-topology, with the sensor nodes representing the edges and the coordinator unit the center of the star. Direct node-to-node communication is also possible, although the predefined processing functionalities do not need it. It is worth noting that SPINE devices communicate atop an application-level protocol, so it is in principle possible to use a multi-hop network layer to realize systems that are based on a physical network in which the coordinator and the nodes are more than one hop distant.

In the following, we describe the software architecture of SPINE, its High-Level Data Processing module, and finally discuss its heterogeneous support of sensor and coordinator device platforms.

3.4.1 Architecture

A high-level representation of the SPINE architecture is shown in Figure 3.1. The SPINE MW is partially located at the coordinator device and partially on the wearable sensors. The MW provides an API both on the coordinator and the sensor nodes to support the development of applications that finally rely on the platform-independent communication protocol layer. This protocol represents an abstraction layer including diverse platform-dependent communication adapters that are dynamically loaded at the coordinator, whereas are linked at compile time at the sensor-node level.

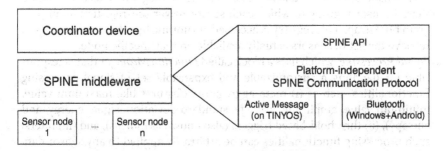

Figure 3.1 The SPINE middleware architecture.

Figures 3.2 and 3.3 show, respectively, the architecture of the SPINE Node(s) and SPINE Coordinator components. The former is implemented in the sensor platform-specific embedded programming language and is placed on each BSN sensor node; the latter is implemented in Java and runs on the coordinator device (an Android porting of the SPINE Coordinator has also been realized).

The *SPINE Node* (see Figure 3.2) consists of four main components:

- *Sensor Node Manager*, which handles the interactions among the Sensing Management, Signal Processing, and Communication modules; it dispatches the requests from the remote coordinator to the appropriate module.
- *Communication*, which handles message reception/transmission and controls radio duty-cycling. It consists of inbound packet decoders (i.e. service discovery, start and reset computation requests, setup function request, function (de)activation request, and setup sensor request) and outbound packets encoders (i.e. service advertisement, buffered sensor readings, processed data message, and acknowledgment packet). Any packet is initially handled by the Radio Controller module, which provides a generic interface independently from the specific underlying radio chip adapter.
- *Sensing Management* (or *SensorBoard* controller), which is the component providing a generic interface to the physical sensors available on the node. It allows to perform one-shot sensor readings and to setup timers for periodic sensor sampling. This component provides easy hardware-independent access to all the supported sensor drivers (SPINE currently supports 3D accelerometer, 2D gyroscope, 4-leads ECG, respiration rate, GSR, EMG, visible and infrared light, humidity, and environmental temperature) through a list of parameterized *Sensor* interfaces. This design choice is motivated by the need for high modularity and efficient customization to support heterogeneous sensing resources in a convenient way. Sensor readings are stored in the *BufferPool*, a data structure that is shared with the Signal Processing module. The BufferPool, internally organized as multiple circular buffers, provides two mechanisms to access the sensor data: (i) upon requests, using getter functions, and (ii) through event listeners that must be registered by interested components (e.g. Signal Processing module) to be notified when new sensor data are available. The Sensing Management also features a shared sensor registry to which each sensor driver self-registers upon program bootstrap. This registry is accessed at runtime by other components to retrieve the list of sensors actually available on that specific node.
- *Signal Processing*, which uses a block called *Function Manager* that is responsible for handling a customizable and expansible set of signal-processing functionalities such as (i) math aggregators (features like maximum value, minimum value, amplitude, average, standard deviation, signal energy, and entropy), (ii) threshold-based triggers (also known as *alarms*), and (iii) filters; such processing functionalities can be arbitrarily applied to any sensor data stream. The Function Manager engine uses an efficient design approach

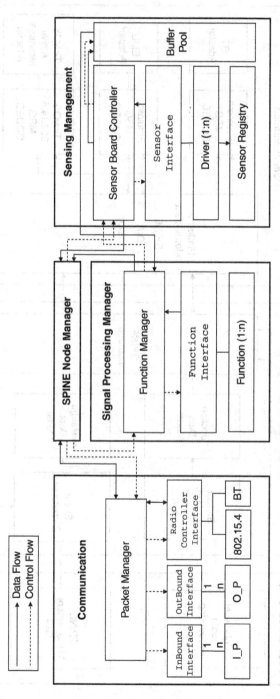

Figure 3.2 The SPINE *Node* software architecture. (Diagram source in Ref. [2]).

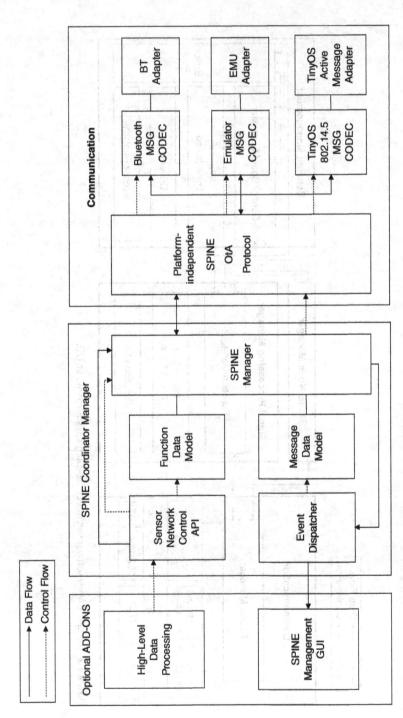

Figure 3.3 The SPINE *Coordinator* software architecture. (Diagram source in Ref. [2]).

based on a list of parameterized *Function* interfaces providing a general-purpose abstraction for any type of processing task. The Signal Processing module retrieves sensor data from the BufferPool and, by interacting with the *Sensor Node Manager* and the *Packet Manager*, it communicates the results to the coordinator unit.

The *SPINE Coordinator* (see Figure 3.3) consists of two main components:

- *Communication*, which has similar functionalities to its corresponding component on the sensor node; it loads at runtime the proper radio module adapter according to the required network stack. It abstracts the logical interactions between the coordinator and the sensor nodes from the actual network activity that depends on the selected platform. This abstraction layer is realized by decoupling the communication interface from its platform-dependent implementation layer.
- *SPINE Coordinator Manager*, which is the most superficial layer atop which every SPINE applications will rely. It is composed of *Sensor Network Control API* (see Table 3.1) and *Event Dispatcher*. The former is an interface used by end-user application developers for the management of the underlying BSN (e.g. to configure the sensors and enable on-node signal processing). The latter is responsible for dispatching events, such as new node discovery and data message arrival, to the registered listeners implemented by the SPINE applications.

3.4.2 Programming Perspective

From a programming perspective, SPINE provides an intuitive Java API (documented in Chapter 12) for convenient BSN management to easily support node discovery, sensing operations, signal processing, and data communication. In addition to several sensor natively supported and pre-defined processing functions, SPINE is designed such that framework tailoring (i.e. customization and extension) becomes very straightforward.

3.4.3 Optional SPINE Modules

The SPINE MW is completed with "optional add-on" modules available only on the coordinator node; they represent an important aspect, despite not being part of its core architecture:

- *High-Level Data Processing*, which provides advanced signal processing and pattern recognition functionalities. It supports the design and implementation of complex applications by means of highly generalized interfaces for data preprocessing, feature extraction and selection, signal processing, and pattern classification. It supports the integration of SPINE in analysis and data mining environments with functionalities such as automatic network

Table 3.1 API exposed by SPINE at the coordinator station.

Functionality	Description
discoveryBsn	Inquiry node discovery and supported sensing and processing capabilities
setupSensor	Allows individual specification of sampling rates for multiple sensors
setupFunction	Setup a preliminary configuration of available processing functionalities
activateFunction	Enables the execution of one or multiple in-node (periodic or trigger based) signal-processing functionalities
startBsn	Issues a broadcast message to the BSN to command a synchronized start of sensing and processing functionalities that have been previously setup and enabled
resetBsn	Issues a broadcast message to the BSN to command a synchronized reset of the nodes
Event	**Description**
newNodeDiscovered	Registered SPINE listeners are notified when a new BSN node is discovered
discoveryCompleted	Registered SPINE listeners are notified when the BSN discovery procedure is terminated
dataReceived	Registered SPINE listeners are notified when new user data sent from a specified node are received by the coordinator
serviceMessageReceived	Registered SPINE listeners are notified when a service message (e.g. warning or error notifications) sent by a specific node is received by the coordinator

configuration and aggregate data collection. It includes a predefined bridge to WEKA [7] (an open-source Data Mining toolkit) to allow the use of its powerful algorithms directly within SPINE.

- *SPINE Management GUI*, which consists of a visual programming tool to configure a SPINE-based BSN without manually coding. According to our experience, it has been useful during initial system testing. Screenshots of its PC and Android implementations are shown in Figures 3.4 and 3.5, respectively.

3.4.4 High-Level Data Processing

The High-Level Data Processing module is an optional SPINE plug-in that empowers the core framework functionalities with additional signal processing and decision-support algorithms (e.g. signal filters, pattern recognition, classification, etc.). This module is available at the coordinator level and

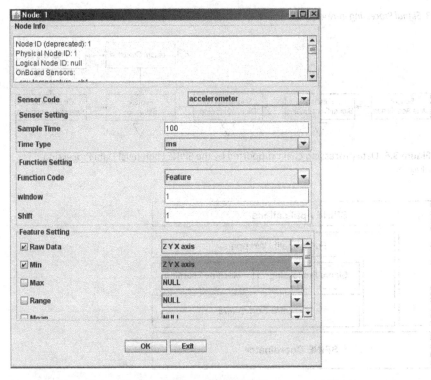

Figure 3.4 *Java desktop* implementation of the SPINE Management GUI (sensor-node configuration dialog window).

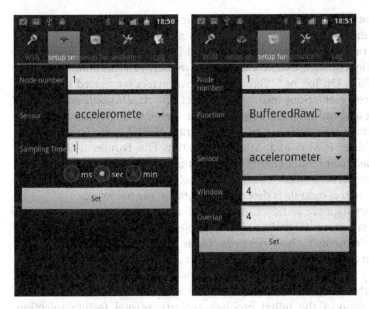

Figure 3.5 *Android* implementation of the SPINE Management GUI (sensor and function configuration dialog windows).

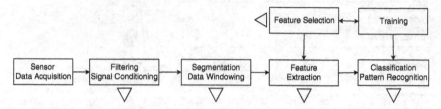

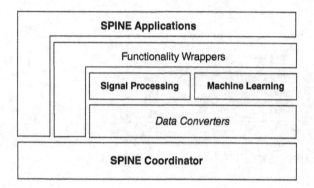

Figure 3.6 Data processing chain supported by the SPINE High-level Data Processing plug-in.

Figure 3.7 High-Level Data Processing layered software architecture.

provides robust support throughout the typical signal-processing workflow, from sensor data acquisition up to classification (see Figure 3.6).

A layered representation of the High-Level Data Processing component is depicted in Figure 3.7. SPINE acts as a MW layer between this module and the underlying BSN. On top of SPINE, a set of converters are placed to convert SPINE data representations into more abstract objects, *Datasets* and *Signals*. Data mining and machine learning tools can therefore transparently handle BSN data, since the module can also generate WEKA-compliant Comma Separated Values (CSVs) and Attribute-Relation File Format (ARFF) files. Finally, a collection of functionality wrappers further support rapid implementation of common tasks needed during the development of SPINE applications. A typical use of this module is described in detail in the following.

BSN sensory data are retrieved with SPINE and converted into more convenient data structures (Signal and Dataset objects, depending on application-specific requirements). Then, developers can optionally apply filtering and segmentation to incoming signals. Feature Extraction algorithms are also available and they are useful when in-node feature extraction functionalities provided by SPINE are not enabled (i.e. SPINE is used to acquire raw sensor signals). To support the initial problem analysis, several feature selection algorithms are provided to identify the most significant subset of extracted features to reach satisfactory classification accuracy. Finally, the classification phase is widely supported, including training. A few algorithms are implemented and ready-to-use; in addition, developers may easily integrate further classifiers, especially thanks to the choice of providing support for using WEKA libraries.

3.4.5 Multiplatform Support

SPINE supports a heterogeneous plethora of hardware platforms, sensors, programming languages, and operating systems; these make this framework suitable for diverse application scenarios (such as smart-Health and e-Fitness), in which, due to specific requirements, only certain hardware/software sensor platforms might be used.

SPINE supports the most common sensor motes. The TinyOS implementation runs on MicaZ, TelosB, and Shimmer/2/2R [8] (for the latter, SPINE supports both IEEE 802.15.4 and Bluetooth radios). This implementation includes a security function using hardware AES-128 encryption of the CC2420 radio. In addition, there exist SPINE implementations for ZigBee devices (like the Telecom Italia "Bollino", equipped with the CC2530 System-on-Chip) based on Texas Instruments Z-Stack and for Java-based Sun SPOT nodes [9]. SPINE also notably provides native support for several physical sensor transducers, including accelerometers, gyroscopes, electrocardiogram, electro impedance plethysmography, temperature, humidity, and light.

In addition to sensors and platforms supported by default, SPINE is designed in such a way that it is easy to integrate further drivers for other sensors and even add support for new platforms. The same happens for the processing functions: integrating additional feature extractors (and even simple classifier algorithms) is straightforward.

At the coordinator level, SPINE supports heterogeneous mobile and desktop devices, as depicted in Table 3.2. Originally, Windows- and Linux-based computers were supported through the SPINE Java SE implementation. However, with the spread of smartphones and tablets having more than sufficient computation and storage capabilities to support mobile-health applications and (almost) continuous Internet connectivity (through which it is possible to transmit raw signals and high-level information to remote servers or in the cloud), we put significant efforts to obtain mobile SPINE coordinators since their use is particularly useful (sometimes strictly necessary) when continuous, outdoor mobility is required and is not possible to rely on fixed infrastructures. A JavaME porting of the framework has been in fact realized. A limited QT implementation is also available and runs on Symbian and Windows smartphones, enabling Bluetooth communication with Shimmer nodes using the third-party QBluetooth library. Finally, and most significant, an Android implementation

Table 3.2 SPINE-tested mobile personal devices.

Device	CPU	RAM (MB)	Miscellaneous
HTC Nexus One	1 GHz, Snapdragon QSD 8250	512	Android 2.x., MicroSD, up to 32 GB
Samsung Galaxy S	1 GHz, ARM Cortex-A8 Dual-Core	512	Android 2.x., MicroSD, up to 32 GB
Samsung Galaxy S4	1.9 GHz, Snapdragon 600 Quad-Core	2048	Android 4.4.2., MicroSD, up to 64 GB
Huawei P8	Quad-core 2.0 GHz Cortex-A53e+Quad-core 1.5 GHz Cortex-A53	3096	Android 6.0, MicroSD, up to 128 GB
Samsung Tab2 10.1	1.0 GHz, ARM Cortex A9 Dual-Core	1024	Android 4.0.3., MicroSD, up to 32 GB
Samsung Note3	2.3 GHz, Snapdragon 800 Quad-Core	3096	Android 4.4.2., MicroSD, up to 64 GB
Nokia N95	332 MHz, TI OMAP 2420 (ARM11-based)	128	Symbian OS v9.2, S60 rel. 3., MicroSD, up to 32 GB
Nokia 6120	369 MHz, ARM11	64	Symbian OS v9.2, S60 rel. 3.1., MicroSD, up to 8 GB

of SPINE has been more recently developed. SPINE Android has been evaluated on several devices (that have been connected to Shimmer nodes over Bluetooth).

Finally, SPINE provides a Java-based emulation environment that virtualizes generic sensor nodes. With this tool, it is possible to emulate a SPINE-based BSN, provided that a dataset is available for each node. Hence, each emulated node is equipped with emulated sensors defined by its given dataset. The SPINE emulator is helpful in various situations; for example, to simplify testing and debugging, processing functionalities can be initially implemented in the emulated environment. In addition, the emulator, along with a simple dataset, has been released in open-source to allow interested developers for investigating the potential of the SPINE framework itself, even if they are not equipped with real wireless sensor nodes.

3.5 Summary

In this chapter, SPINE, a domain-specific programming framework, has been presented. The main goal of SPINE is to provide BSN developers with support for rapid prototyping of signal-processing applications. In SPINE, sensors and common processing functionalities, such as math aggregators and threshold-based alarms, can be configured independently and connected together arbitrarily at runtime based on external controls.

Hence, one of the main achievements of SPINE is the reuse of software components to allow different end-user applications to configure sensor nodes at runtime based on the application-specific requirements without off-line reprogramming when switching from an application to another. Furthermore, thanks to its modular component-based design approach, SPINE enables a great degree of heterogeneity: a wide variety of hardware platforms, sensors, programming languages, and operating systems are supported. This allows for a very flexible and usable framework in different BSN application scenarios, where, due to specific requirements, only certain platforms or operating systems might be used.

References

1 Bellifemine, F., Fortino, G., Giannantonio, R. et al. (2011). SPINE: a domain-specific framework for rapid prototyping of WBSN applications. *Software: Practice & Experience* 41 (3): 237–265.

2 Fortino, G., Giannantonio, R., Gravina, R. et al. (2013). Enabling effective programming and flexible management of efficient body sensor network applications. *IEEE Transactions on Human-Machine Systems* 43 (1): 115–133.

3 SPINE Website. http://spine.deis.unical.it (accessed 8 June 2017).

4 Tinyos Website. www.tinyos.net (accessed 14 June 2017).

5 Gay, D., Levis, P., von Behren, R. et al. (2003). The NesC language: a holistic approach to networked embedded systems. *ACM SIGPLAN Notices* 38 (5): 1–11.

6 Von Eicken, T., Culler, D., Goldstein, S.-C., and Schauser, K.-E. (1992). Active messages: a mechanism for integrated communication and computation. *Proceedings of the 19th Annual International Symposium on Computer Architecture, ISCA'92*, Queensland, Australia (19–21 May 1992), pp. 256–266. ACM Press.

7 Holmes, G., Donkin, A., and Witten, I., Weka: a machine learning workbench. *Proceedings of the 2nd Australia and New Zealand Conference on Intelligent Information Systems, ANZIIS'94*, Brisbane, Australia (29 November–2 December 1994), pp. 1269–1277. IEEE Press.

8 Shimmer Website. www.shimmersensing.com (accessed 5 June 2017).

9 SunSPOT Website. www.sunspotdev.org (accessed 10 June 2017).

4

Task-Oriented Programming in BSNs

4.1 Introduction

The SPINE framework described in Chapter 3 provides an effective solution for easily and rapidly developing highly customizable signal-processing applications for BSNs. The in-node processing applications supported by SPINE are usually defined as a three-layer chain of tasks: (i) acquisition of raw data streams from the sensors, (ii) computation of processing functions on the data streams to extract specific features, and (iii) transmission of processed data to the base station for further computation.

However, some signal-processing applications require an extension of this approach to fully satisfy the needs for a more complex composition of sensing and processing tasks. Therefore, a *task-centric programming* model has been experimented in a new reengineering of the SPINE framework, dubbed *SPINE2* [1]. Conceived not to be a replacement for SPINE 1.x versions, SPINE2 is actually intended as an alternative application design tool exposing a different methodology to translate the high-level intentions of the developers into actual executable routines to be deployed on a BSN. The task-oriented approach aims at providing an easy and effective way for developing distributed signal-processing applications, thanks to its intuitive and graphical design model. It offers a wide range of benefits to developers, like the advantage of abstracting away low-level details of the sensor platforms and their operating system as well as the complexity of managing the communication among nodes. Moreover, a platform-independent middleware eases the reusability and portability of the code and the interoperability of applications among heterogeneous embedded environments, while not neglecting the stringent requirements in terms of execution efficiency and stability.

In this chapter, the SPINE2 programming paradigm and the software architecture of the underlying distributed middleware running on the sensor nodes are presented. With SPINE2 we show how fairly sophisticated signal-processing applications can be realized in the form of easy-to-implement embedded processes.

Wearable Computing: From Modeling to Implementation of Wearable Systems Based on Body Sensor Networks, First Edition. Giancarlo Fortino, Raffaele Gravina, and Stefano Galzarano.

4.2 Background

The main limitation in developing applications for BSN-based systems is the need for proper design and programming skills to successfully deal with the low-level aspects of embedded devices. Also, application development is even more challenging and time-consuming due to the very resource-constrained environments provided by the most commonly available sensor platforms. Unfortunately, such a hard task prevents BSN-domain experts, who may not have a software development background, from directly contributing to the building of applications. Therefore, a proper high-level development paradigm is highly desirable to hide the low-level programming issues, so as to allow anyone with poor or no skills in programming to autonomously prototype and test their own applications by focusing on the desired algorithms. Such a desired paradigm should come with a set of well-defined constructs that lead to a faster application definition as well as a more component reusability and a minimized maintenance process.

That entails the adoption of abstract, easy-to-use, and fully configurable functional blocks, which should allow to quickly implement the set of the most common operations needed by BSN applications. Interoperability and interconnection among applications, possibly defined by different users, should also be part of the features supported by a high-level development framework. This is usually achieved by defining a common higher level communication protocol, which is independent from the actual underlying protocols supported by a specific sensor platform. Moreover, its paradigm philosophy should strongly promote the potential benefit of enabling an easier application reconfiguration at runtime and thus providing built-in mechanisms for dynamic reprogramming without directly accessing the already deployed devices.

The idea of employing the well-known task-based paradigm in the BSN context comes from the need of finding a better way to meet all these requirements with an easy-to-understand high-level paradigm able to (i) effectively and efficiently abstract away from the hardware and the network-specific details and (ii) provide constructs particularly devoted to easily define distributed signal-processing applications.

4.3 Motivations and Challenges

4.3.1 Need for a Platform-Independent Middleware

Applications' interoperability is completely achieved when their interaction is made possible even in the case of execution over different heterogeneous sensor devices. This implies the need of a programming framework capable of transparently supporting a diversified hardware and software environment. As a consequence, the simplicity in making the entire middleware infrastructure ported to a new sensor platform is a further desirable requirement and of crucial relevance for a more widely use of the framework in complex real-world applications involving heterogeneous computing systems. Differently from a platform-specific software architecture, the desired middleware should not be developed exclusively by using the library provided by a platform-specific programming environment. Conversely, by adopting a more generic programming language for implementing the core functionalities of the middleware in charge of executing the high-level abstractions, this common software layer should be able to run over different platforms (supporting such a generic language) with little or no additional code.

4.3.2 Challenges in Designing a Task-Oriented Framework

In the following, the challenges in developing a framework meeting the aforementioned requirements are discussed. To summarize, it is of crucial importance to keep in mind the following desired requirements while designing a BSN framework/middleware:

- *Proper easy-to-use high-level programming paradigm*: since the adoption of programming methods based on high-level models can greatly improve productivity, a definition of good and easily understandable abstractions to hide low-level platform-specific operations represents the main key factor for the success of a programming framework. In particular, a major challenge is finding a good adjustment of the generic task-based approach for the specific needs in the BSN domains, while fulfilling the nonfunctional requirements (efficiency, portability, and interoperability).
- *Heterogeneity*: the ability to deploy the same applications over different sensor platforms in a transparent way for the developer should also be a must, since it would allow for a holistic approach to managing diverse sensor networks and applications.
- *Portability*: in order to prolong the framework lifetime and keep it up-to-date over the time, the design of the node-side middleware architecture should be properly performed to support a seamless portability process across new sensor platforms and embedded systems. However, this is not a trivial problem.
- *Extensibility*: the middleware should also rely on a modular architecture for an easier introduction of new processes, functionalities, and communication capabilities as well as the integration of new physical sensors and drivers. It is not simple to design a middleware that guarantees an easy update of components.
- *Efficiency*: the aforementioned features would be of little importance with a resource-hungry middleware. Good runtime performance should be achieved despite the stringent resource constraints of the common sensor platforms.

4.4 SPINE2 Overview

The SPINE2 framework has been conceived to further increase simplicity and effectiveness in developing distributed signal-processing application atop BSNs. Specifically, its peculiar feature is the adoption of a *task-oriented paradigm*, which allows developers to quickly make use of simple constructs to translate the high-level application logic (global behavior) into actual operations to be executed on each single sensor node of the networks. Moreover, SPINE2 makes reconfigurability and reusability of applications easier than other proposed programming frameworks for sensor networks.

SPINE2 comes with two main software components: the sensor node middleware running on the network and the management software running on the coordinator side (typically a PC or a supported hand-handled device). The latter one, developed in Java, is the main interface to the BSN. In particular, it provides well-defined APIs, thanks to which developers can easily manage the network as well as the application, i.e. defining, deploying, and running the defined set of interconnected tasks. Moreover, it gathers the data preprocessed on the nodes, which can be further processed by more complex and resource-demanding user-defined algorithms and visualization tools. The node-side middleware running on top of the sensor node operating system has two main functions: (i) handling messages received from the coordinator or any other node and (ii) managing and executing the tasks the node is responsible for.

The key characteristics of the framework are discussed below.

- *Platform independence and quick portability*: supporting a swift portability across diverse sensor platforms was one of the primary motivations for which SPINE2 has been designed. As such, the node-side middleware architecture has been conceived for decoupling the task runtime execution engine from any other services provided by specific operating systems, as depicted in Figure 4.1. Based on the *software layering* approach, the whole runtime system of the node is composed of two main set of components. The "core modules," which are implemented in C language, is developed to support any C-like sensor platform with any or little need for modifications. Underneath the core, a set of "platform-specific modules" are properly defined as adaptors for allowing the core to interact with the operating system services and resources (sensors, timers, communication, etc.). Different adaptors interface with specific sensor platforms and software environments, such as TinyOS [2] and Z-Stack [3] (the ZigBee-compliant implementation provided by Texas Instruments). The benefit of such an architecture is that a developer needs to implement just the necessary adaptation modules in order to deploy the platform-independent components and the applications onto new sensor platforms.

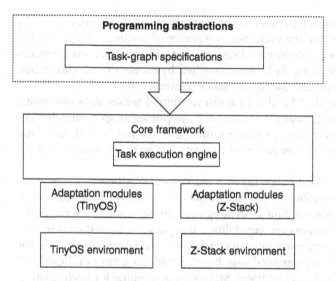

Figure 4.1 The software layering approach in the SPINE2 middleware.

- *Extensibility and customization*: thanks to the task-oriented approach, it is possible to easily add new functionality when the need arises. This is done by defining a new task implementing a user-defined computing logic without having to change the underlying runtime environment. New drivers for sensors or actuators can also be added by simply developing proper adaptation modules.
- *Modularity*: the node-side middleware architecture, described in Section 4.6, includes independent modules interacting through well-defined interfaces, with the benefit of easier software maintenance and upgrade processes.

4.5 Task-Oriented Programming in SPINE2

The task-oriented programming paradigm provided by SPINE2 is specifically conceived to support the creation of data-flow-based task chains for defining distributed signal-processing applications. Less error-prone with respect to explicitly coding a low-level code, this approach is more intuitive as the user, according to the application requirements, has to specify a set of interconnected tasks made available from a task library. Thus, the basic abstract components constituting the high-level application model are *tasks* and *task-connections*.

A task represents a specific activity or operation, e.g. a signal-processing function, a data transmission, or a sensor querying. Tasks are executed in an atomic way with respect to other tasks, whereas they can be interrupted by triggered events. In fact, the event-reactive nature of the sensor nodes implies the need for a fast response to asynchronous events like a radio message reception or a timer expiration. Tasks are connected by means of task-connection representing temporal and data dependency between tasks.

In Figure 4.2, a typical (in this case rather simple) sensor data-processing application is shown. It basically consists of three phases: (i) gathering the sensor readings, (ii) executing processing functions on the sensed data, and (iii) sending results to other nodes of the network or to the coordinator for further elaboration.

In order to achieve load-balancing of resources and an efficient communication, SPINE2 allows to allocate specific subset of tasks to different nodes, thus realizing a full distributed data processing over the network. Since the nodes may have different features and capabilities, it is possible, for instance, to allocate the most computational-intensive tasks to more powerful nodes in the network. Thus, the implemented task-based paradigm gives developers full control over data feed, control flows, and event scheduling for performance-balancing on multiple dimensions (e.g. CPU, memory, and energy). Moreover, composing an application by means of basic functional blocks with well-specified inter-task interface allows an easy and rapid application reconfiguration and a simpler maintenance process. The library of reusable tasks includes two main types of tasks:

- *Functional tasks*: perform data processing/manipulation or execution control.
- *Data-routing tasks*: provide data forwarding or replication.

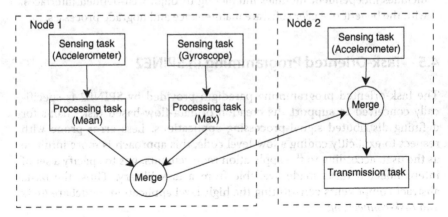

Figure 4.2 A task-oriented application with tasks instantiated on different nodes.

Each task is defined as a triple of attributes: *Input, Output,* and *Parameters*. Depending on the specific functionality for which a task has been defined and implemented, the user can configure it by means of a set of pairs (*parameter* and *value*). Also, there may be zero to multiple (usually in data-routing tasks) input or output connections. Each connection can handle a particular sensed data, processed information, or even an *empty* data, intended to be used as a simple "execution complete notification" for the connected task.

The main tasks constituting the currently available library are the following:

- *TimingTask*: defines timer expiration and can be adopted for timing other tasks. It has no input connection and does not have to process any data. It signals a notification through its output when its inner timer expires depending on the following parameters: the *periodicity* (i.e. specifies if it is a periodic timing or a one-shot expiration), the *period of expiration*, and the corresponding time *scale/unit*.

- *SensingTask*: performs a reading from a particular onboard physical sensor. It encloses an inner timer for scheduling the sensing operation. The data output depends on the specific type of sensor it is configured to read data from. Specifically, it can consist of a simple scalar reading value (e.g. when linked to the luminosity sensor) or a vector of samples each coming from a specific "sensor channel" (e.g. a triaxial accelerometer provides three different samples).

- *ProcessingTask*: provides the actual computing capabilities by performing functions or algorithms to process data. Some set of functions are called "feature extractors," which are usually applied to temporal data series. Some examples are mean, variance, max, and min.

- *TransmissionTask*: is in charge of explicitly transmitting data coming from other connected tasks to a specific destination node/device. It is usually used for sending in-network preprocessed data to the coordinator of the BSN. In the case of interconnected tasks deployed on different nodes, the SPINE2 middleware performs proper data transmission (encapsulated in proper messages) without the need of a TransmissionTask.

- *StoringTask* and *LoadingTask*: perform data (stream) storage and retrieval by using the onboard flash memory, if available on the platform.

- *SplitTask*: duplicates incoming data from its input connection to all its output connections, so as to make it available to multiple tasks.

- *MergeTask*: merges incoming data from its input connections and feeds its single output connection; it first normalizes and/or uniformly formats the collected data.

- *HistoricalMergeTask*: performs a number (specified by a parameter) of sequential merge operations over the time and makes the collected data available to the output.

4.6 SPINE2 Node-Side Middleware

The main purpose of the middleware running on the nodes of a SPINE2 sensor network is to "interpret" and "execute" the high-level application defined through the task-oriented paradigm. Figure 4.3 depicts its modular architecture composed of a set of modules, each including interacting (but independent) software components intended to accomplish well-defined operations.

The *core framework* of SPINE2 (see also Figure 4.1) is made up of all the components in white blocks of Figure 4.3. Implemented in ANSI C language, they can be compiled in any "C-like" development environment with no changes in their inner code. The core encloses all the unchangeable parts of the middleware implementing the main runtime task execution logic, including task and memory management, application-level message handling, and abstract access to onboard sensors and actuators. By contrast, the grey blocks are the *architecture-dependent* part of the middleware and are tailored for a specific sensor platform in order to manage the lower level mechanisms and services. Some adaptation components (or drivers) bridge the core with the platform by granting access to the physical resources through well-defined interfaces.

The use of a common programming language, and its standard libraries, along with a strong software decoupling between the core and the platform-related components are the key characteristics for the very high portability of the SPINE2 middleware.

A more specific description of the modules shown in Figure 4.3 is provided below.

- *SPINE2Manager*: is the central component of the architecture. Its main functionalities include (i) system initializing at startup, (ii) orchestrating the modules managing the node resources (sensors, actuators, radio, and flash memory), (iii) dispatching the necessary commands to the other components to accomplish required operations (e.g. a new task creation or a buffer allocation), and (iv) handling the SPINE2 application-level protocol (see Section 4.7) for communication with the coordinator and the other nodes, which includes formatting of the SPINE2 outgoing messages before its encapsulation into a low-level packet by the *Comm Module*.
- *Comm Module*: provides the basic services for exchanging messages with the other sensor nodes and the BSN coordinator. It encapsulates the application-level messages into packets and performs the reverse operation, by also handling the (de)fragmentation operations when required, depending on the message length and on the maximum payload supported by the platform-specific communication protocol (see Section 4.7).
- *Task Module*: is the middleware "task execution engine" in charge of (i) instantiating the tasks allocated on the node by the coordinator, (ii) scheduling, and (iii) terminating their executing based on the inter-task connections.
- *Memory Module*: handles the memory space by allocating the task-based application definitions as well as the buffers required both for the inter-task

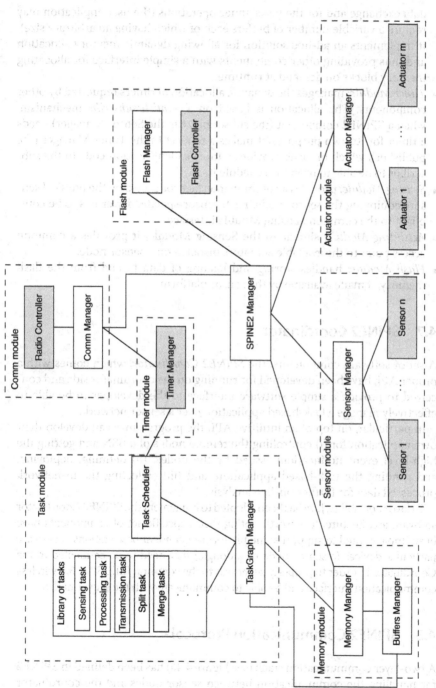

Figure 4.3 Software architecture of the node-side part of the framework.

data exchange and for the tasks' inner operations (the user application may require a variable number of buffers each of which having an arbitrary size). It implements an ad-hoc solution for allowing dynamic memory allocation and thus providing other components with a simple interface for allocating memory blocks on demand at runtime.

- *Timer Module*: manages the dynamic allocation of timers requested by other components. The allocation is based on a *publish/subscribe* mechanism: when a SPINE2 component (the subscriber, e.g. the Sensor Manager) needs a timer for its own purposes, it makes a request to the Timer Manager (the publisher), which in turns provides a timer's identification code to the subscriber to be able to properly schedule it.
- *Sensing Module*: provides a common interface for accessing the physical sensors equipping the sensor node. Each sensor-specific driver has to be compliant to the common Sensing Module interface.
- *Actuating Module*: similar to the Sensing Module, it provides a common access point to the available actuators installed on a sensor node.
- *Flash Module*: handles storing and loading of data to and from the flash memory, if made available by the sensor platform.

4.7 SPINE2 Coordinator

A set of software components, the *SPINE2 Coordinator*, which comes with a proper API, have been developed for running on the coordinator side and conceived to provide a simple software interface to the developer to be able to effectively manage a task-based application over a sensor network.

In particular, on top of an intuitive API, the programmer can develop their own application for (i) controlling the remote nodes of a BSN and getting the high-level event notifications issued by the nodes, (ii) defining, deploying, and running the task-based application, and (iii) collecting the in-network processed data for further off-line analysis.

To favor portability, Java has been adopted to implement the SPINE2 Coordinator software architecture. It is worth noting that a specific set of components have been implemented to support some platform-dependent base stations. These are particular devices (a sensor node or a dongle) that need to be connected to the Coordinator in order to properly get access to the common IEEE 802.15.4 wireless communication interface and be able to communicate with the sensor nodes.

4.8 SPINE2 Communication Protocol

A two-layer communication stack (see Figure 4.4a) has been defined in SPINE2 for handling the communication between sensor nodes and the coordinator and is built atop the platform-specific protocol supporting the available

(a)   (b)

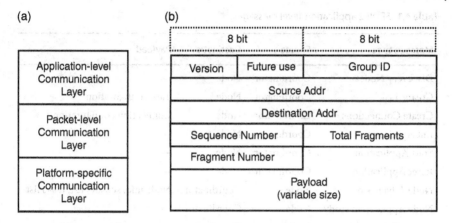

Figure 4.4 The two-layer protocol stack (a) and the packet fields (b).

onboard radio. The layer in the middle (*Packet Layer*) provides a point-to-point communication interface, by also managing the fragmentation of the long application-level messages (split into multiple packets' payload) coming from the upper layer. In particular, the fields constituting the SPINE2 packet are depicted in Figure 4.4b.

The upper layer is defined to handle a set of SPINE2 messages, which encapsulate the application-level commands and information for interacting with the BSN and more specifically with the deployed task-oriented application.

The currently supported application-level messages are summarized in Table 4.1 along with some additional information about the communication direction and the carried payload. The *Init Application, Start Application,* and *Reset Application* Messages, which have no additional payload data, are adopted for controlling the execution of the task-based application once correctly deployed.

The *Discovery Nodes* initiates the communication scheme between the coordinator and the BSN in order to get general information from the nodes (through the *Node Advertisement* message), like the sensor platform, the available onboard sensors, and the list of supported tasks. Once the Discovery/ Advertisement phase has terminated, the user can complete modeling the application, which is then deployed by mapping the task-graph throughout the network. The *Create Task* Message is issued to instantiate each single task on the intended node. Similarly, the *Create Connections* message is sent to create a connection or a set of connections between tasks. It therefore includes information related to the destination task of a specific connection since a task may be either local (i.e. instantiated on the same node) or remote. It also includes information for allocating the needed buffers on the node. Once the application has been deployed, the coordinator can broadcast the *Init Application* to initialize the tasks instantiated over the network, after which

Table 4.1 SPINE2 application-level messages.

Message type	Source	Destination	Payload
Discovery Nodes	Coordinator	Node	—
Create Task	Coordinator	Node	Task configuration
Create Connections	Coordinator	Node	Connection configuration
Init Application	Coordinator	Node	—
Start Application	Coordinator	Node	—
Reset Application	Coordinator	Node	—
Node Advertising	Node	Coordinator	Node info, sensors list, tasks list
Node Application ready	Node	Coordinator	—
Sensor Data	Node	Coordinator	Formatted data
Error	Node	Coordinator	Error code, error info
Status Info	Node	Coordinator	Status code, status info
Sensor to Sensor Data	Node	Node	Formatted data

every node communicates that it is ready to run (part of) the application by sending a *Node Application Ready*. The *Start Application* Message can now be broadcast, causing the application to run. The *Sensor Data* message is for forwarding data (either raw or preprocessed) from a node to the coordinator, whereas the *Sensor to Sensor Data* message is for data that needs to be exchanged between remote tasks. The *Error* and *Status Info* messages are issued in case of unexpected errors at runtime (e.g. no further block can be allocated in the dynamic memory) or for periodic node status advertisement (e.g. for communicating the remaining battery charge).

4.9 Developing Application in SPINE2

A typical interaction between the SPINE2 environment and the Java-based user-defined applications is depicted in Figure 4.5. It is worth noting that a SPINE2 Console is made available along with the SPINE2 Coordinator component. Specifically, it comes with a simple GUI that allows a user to immediately interact with the BSN and define the task-based application without having to implement an application atop the SPINE2 API.

As a consequence of the presence of such GUI, a developer can actually interface his own applications to the SPINE2 environment in two different ways.

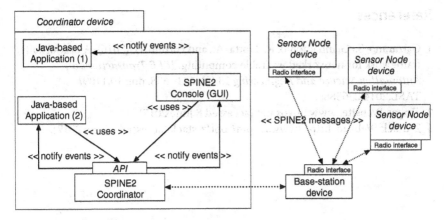

Figure 4.5 The SPINE2 components interacting with the user applications.

As a first example, since the SPINE2 Console can be adopted for managing both the BSN and defining the task-based application, the Java-based *Application 1* needs just to register to the Console in order to get notified of messages coming from the network, thus needing only to code the logic for collecting information and data useful for off-line analysis and displaying.

By contrast, *Application 2* directly makes use of the SPINE2 API and takes care of both managing the BSN and the task-based application, thus requiring a greater effort by the developer.

4.10 Summary

This chapter has presented the SPINE2 programming framework, an easy-to-use solution for rapidly and effectively developing distributed application on BSNs. We have first provided motivations for needing different programming abstractions, by explaining why the well-known task-based paradigm is able to successfully meet the desired requirements of the BSN domains. Then, an overview of the main features of SPINE2 has been presented, along with a description of the supported task-oriented programming approach and related benefits: rapid prototyping and easy runtime reconfiguration of highly customizable and flexible distributed signal-processing applications. Furthermore, its software architecture has been described by highlighting the benefits in having a platform-independent node-side middleware in terms of quick portability and extensibility.

References

1 Galzarano, S., Giannantonio, R., Liotta, A., and Fortino, G. (2016). A task-oriented framework for networked wearable computing. *IEEE Transactions on Automation Science and Engineering* 13 (2): 621–638. doi: 10.1109/TASE.2014.2365880.

2 Tinyos Website. www.tinyos.net (accessed 8 June 2017).

3 Z-Stack Website. http://www.ti.com/tool/z-stack (accessed 5 June 2017).

5

Autonomic Body Sensor Networks

5.1 Introduction

High-impact applications enabled by BSN-based systems are required to be secure, safe, and reliable, especially when dealing with the monitoring and controlling of the physical and biochemical parameters of the human body. Achieving correctness, accuracy, and efficiency at execution time by meeting the strict requirements in terms of fault tolerance, adaptability, and reliability is of crucial importance and a very challenging issue. In this regard, the autonomic computing paradigm can perfectly fulfill such critical requirements of BSN applications in which proper techniques can be incorporated to enable specific self-managing capabilities and successfully cope with unforeseen changing conditions that may lead to unpredictable behaviors.

This chapter first introduces background concepts on the autonomic paradigm and its application on the BSN context. Then, the needs for BSN-specific autonomic-enabling development tools are discussed. Finally, a framework conceived to support rapid design and implementation of applications having autonomic properties, *SPINE-**, is presented. Implemented as an extension of SPINE2, the autonomic elements are incorporated into the same high-level abstractions adopted for developing the BSN applications. Specifically, it aims at easily integrating the autonomic behavior without affecting the applications, thanks to the adopted task-oriented paradigm, which allows for the required separation of concerns between the user-defined application business logic and the autonomic-related operations.

5.2 Background

The term Autonomic Computing (AC) was coined by researchers in IBM [1], who advised the need for a management component acting in a similar fashion to the autonomic nervous system of the human body, in response to the

Wearable Computing: From Modeling to Implementation of Wearable Systems Based on Body Sensor Networks, First Edition. Giancarlo Fortino, Raffaele Gravina, and Stefano Galzarano.
© 2018 John Wiley & Sons, Inc. Published 2018 by John Wiley & Son, Inc.

increasingly complexity of managing computing systems. The AC paradigm was then conceived for dealing with the complexity of distributed software systems and enabling mission-critical applications to meet high reliability and adaptability requirements. It faces the problem by introducing a series of self-* properties, thanks to systems that are able to perform several self-management actions with no direct human intervention. The main self-* properties (usually known as self-CHOP properties) are the following:

- *Self-configuration*: depending on high-level policies and objectives, a system is able to effectively configure and adapt itself on the basis of the user's needs and environmental conditions by dynamically adding, replacing, or removing its components with no system outages.
- *Self-healing*: to guarantee an adequate level of reliability, the system should autonomously prevent, detect, and possibly remedy malfunctions and errors. The nature of possible problems that can be detected spans from low-level hardware failures to high-level erroneous software configuration. However, it is important that the operations related to the self-healing process do not affect other vital components in the system.
- *Self-optimization*: the system should perform its activities by proactively and effectively targeting the maximum performance given the restricted available resources. This optimization process should constantly seek performance improvement without interfering with the system in achieving the user-defined goals.
- *Self-protection*: systems with such a property are able to guarantee an adequate level of security in terms of detecting, and possibly preventing, malicious attacks aimed at disrupting the normal planned system operations. Moreover, the system should also protect itself from user inputs that may be inconsistent, implausible, and dangerous.

5.3 Motivations and Challenges

As discussed in the previous chapters, BSN developers can benefit from the use of programming frameworks (e.g. SPINE, SPINE2, and MAPS), which target ease of development, fast prototyping, code reusability, efficiency, and application interoperability. However, the global quality of the applications not only derives from the use of a well-defined programming approach and related tools, but also on how good they are designed and implemented to deal with the changing conditions and possible problems due to the interaction with the environment and other interconnected systems. In fact, since unpredictable conditions (e.g. sensing faults) may lead to unwanted behaviors at execution time, it is not reasonable for a BSN system to be constantly supervised and maintained by human operators once deployed. Therefore, despite usual development issues are proven to be successfully addressed by the most common programming frameworks, the way in which the correctness of applications, during the post-deployment stage, has to be defined is usually completely up to the developers. And this is becoming a particularly challenging task in view of the fact that evermore complex BSN applications will need a better runtime support as a result of the immersion of people into more pervasive, smarter, but also risky environments.

Providing an effective approach to allow developers providing self-managing capabilities and easily integrating them into applications, in order to improve reliability and maintainability, is a major challenge. Unfortunately, most of the currently available BSN programming frameworks represent trustworthy tools for defining the high-level application logic, but they do not provide an explicit and clear way for designing an underlying autonomic structure capable of addressing the application management requirements.

5.4 State-of-the-Art

The integration of the autonomic principles into networking systems has been studied and proposed in many research works [2, 3]. Also, real prototypes have been developed, deployed, and tested as releases of several international projects: BISON [4], ANA [5], Haggle [6], CASCADAS [7], EFIPSANS [8], and Autonomic Internet [9].

However, differently from traditional networks, the peculiar characteristics of sensor networks make the design and implementation of the autonomic management approaches even more challenging, and to date, this branch of research has not been satisfactorily investigated yet. Examples of autonomic-oriented system architectures explicitly designed to support sensor networks management are MANNA [10], BOSS [11], WinMS [12], and Starfish [13].

MANNA [10] is a generic architecture providing three different abstraction planes, one for each management function: functional areas, management levels, and WSN functionalities. The latter includes basic low-level operations like sensing, processing, and communication, whereas the management levels represent the typical system's layers, i.e. the business logic, the middleware services, and the networking layer. Finally, the functional areas represent, for each aforementioned system's layer, the different perspective to which autonomic actions can be applied, specifically configuration, maintenance, performance, security, accounting, and fault management perspective.

Based on the standard UPnP protocol, the BOOS architecture [11] is designed to support automatic discovery, configuration, and controlling of devices over traditional networks, by avoiding any manual setup. Due to the limited resources of sensor devices, to fully support UPnP functionality, a mediator component running on the coordinator serves as a provider of the required services for network management. The BOSS architecture is constituted of several functional components: control manager, service manager, event handlers, and sensor network-level management functions.

WinMS [12] is a network management system able to support dynamic adaptation of nodes as a response to changing network conditions. Depending on high-level policies, WinMS is based on a local management scheme, which works according to the neighborhood network state, and a decentralized scheme, which depends on global network-level knowledge. The low-level communication is provided by a lightweight TDMA protocol, FlexiMAC, supporting a tree-based gathering scheme, which is in charge of collecting and disseminating network state data and management information.

Starfish [13] is a framework conceived to support the definition of self-adaptive behaviors in sensor networks. Specifically, a node-side policy management system, called Finger2, is in charge of executing the adaptive strategies dealing with the management of reconfigurations and failures. Such strategies are specified by the developers through a desktop client tool, which includes a set of libraries to facilitate the programming of nodes by providing a high-level language for defining both autonomic policies and user's application logic.

Although the previously described frameworks and architectures are examples of generic self-management systems, most of the current research efforts are mainly focused on self-healing and fault management [14–24]. Moreover, these studies are usually carried out by considering the WSN context, whereas few efforts are devoted to BSNs. This is why we aim at addressing such a shortcoming by specifically exploring the viability and convenience of autonomic computing in the BSN context.

5.5 SPINE-*: Task-Based Autonomic Architecture

Due to the intrinsic complexity of distributed computing systems, like BSNs, there exist different approaches for integrating the autonomic properties, which can be applied at different system's perspectives: network-level, communication stack-level, software layer-level, service-level, function-level, or component-level.

However, we advise the practice of clearly and explicitly separating the application business logic from the implemented autonomic management operations. If well-designed, the main benefit from using such separation of concerns is that the application developer's efforts can be focused on the characteristics of the application and its primary goals, without being forced to take care of any of the autonomic management components. In fact, the autonomic behavior can be easily added afterwards, with no risk of affecting the previously defined application logic.

In the following, an autonomic architecture fulfilling the aforementioned requirements is presented. It has been designed and implemented around the SPINE2 framework, whose task-based abstractions provide the necessary mechanisms for assuring the application isolation and composition properties. The autonomic features have been added to SPINE2 without affecting the original runtime engine but instead only involving its task library, which has been enhanced with the introduction of a new set of autonomic-specific tasks. The way a SPINE-* application can be defined is depicted in Figure 5.1. Such an application is constituted by a multiplane architecture which, in its basic configuration, is composed of two distinct planes, one representing the user application logic and the other providing the autonomic operations. Since a task is only aware of its input data, it is clearly possible to employ a generic nonautonomic task in the autonomic plane on the basis of specific needs. Also, it is worth noting that, differently from the application example provided in Chapter 4 (see Figure 4.2), all tasks have been depicted with no assumption on their specific types.

Different kinds of interactions can be established between the planes for performing direct manipulation on the application data streams or reconfiguring the application tasks. Despite such interactions, the degree of isolation in the execution of tasks guarantees that the separation of concerns' property still holds, with the application plane having no awareness of the presence of the autonomic plane.

The generic architecture of Figure 5.1 shows two specific autonomic approaches. In the first case, the parameters of task T7 are tuned at runtime so as to optimize its function and thus adapting its behavior depending on the

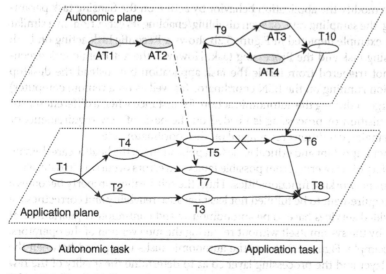

Figure 5.1 The multiplane autonomic architecture of a SPINE-* application.

data originating from task T1. Specifically, the adaptation action is performed by the autonomic task AT2 after a preprocessing made by AT1 on the source data coming from T1. Since no data stream is injected to the application plane, but rather a reconfiguration action is performed, such an interaction (configuration connection) is represented with a dashed arrow from AT2. In the second case, supposing that the aim of the autonomic tasks AT3 and AT4 is to improve the quality of data, the output data stream from T5 is redirected to the autonomic plane, specifically to T9. In turn, T9 provides AT3 and AT4 with the data stream to be analyzed and manipulated prior to feed the aggregator task T10, which is in charge of fusing the two data streams and sending the resulting stream to T6. In such a configuration, the direct connection T5–T6 has been removed and replaced with the subgraph of tasks in the autonomic plane.

The proposed task-based multiplane autonomic architecture can be employed in many common situations in which self-* properties need to be satisfied. In the following, some examples of task-based application enhanced with the SPINE-* autonomic mechanisms are presented. In particular, we show the four self-CHOP properties: self-configuration, self-healing, self-optimization, and self-protection.

As represented in the reference architecture of Figure 5.1, a useful property of BSN applications is the ability of autonomously reconfiguring the parameters of a task at runtime depending on the changing system and/or environmental conditions. As shown in Figure 5.2, two different ways for triggering a reconfiguration task can be adopted. In Figure 5.2a, the SensReconfig task of the autonomic plane is driven by the output results of the Processing task (which performs some kinds of analyses on raw sensed data). Specifically, the SensReconfig task is able to modify the application behavior by acting on the Sensing task parameters, e.g. the sampling rate, or even disabling/enabling its execution. In a similar way, the example depicted in Figure 5.2b shows a ReconfigTask acting on both the Sensing task and the Processing task. However, the autonomic task execution is not triggered from inside the task application but instead the desktop application running on the BSN coordinator (as well as on a remote computer) is in charge of driving the autonomic action, for instance when a different sensor data acquisition or processing is needed on the basis of new requirements or some changing conditions recognized on the coordinator side.

Another important and critical issue for applications in the health-care domain is the ability of recovering from possible faults and errors occurring in data, algorithms, or networking functionalities. Thus, the self-healing property becomes a crucial requirement to be fulfilled not least because reliability and correctness of the provided services have to be autonomously and continuously guaranteed at runtime by the system itself without requiring the intervention of the operators. As an example, Figure 5.3 shows the autonomic tasks interposed between the sensing layer and the processing layer so as to determine the quality of the raw data from the sensors and thus to avoid that corrupted samples (when detectable)

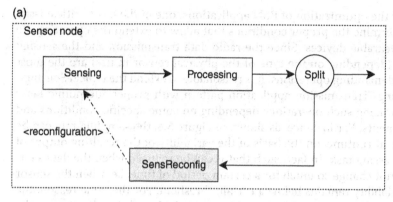

(a)

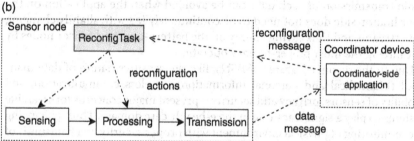

(b)

Figure 5.2 Examples of application with self-configuring property; (a) the reconfiguration task is driven by the output results of the Processing task; (b) the reconfiguration task is driven by the desktop application running on the BSN coordinator.

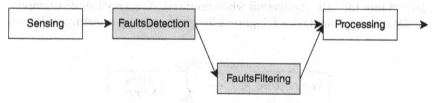

Figure 5.3 Example of application with self-healing property.

could affect the computing functions and thus the whole application accuracy and erroneous behavior. Specifically, the FaultsDetection task may be conceived as an online detection process for specific faults in the data stream coming from the Sensing task, which is also in charge of possibly redirecting the corrupted stream to the FaultsFiltering task for the actual recovering process. As discussed in Ref. [25], different kinds of data faults can heavily impact the application correctness. Moreover, it is also shown that not all of them can easily be treated with proper recovery techniques in order to improve system's tolerance to data faults and then achieve better efficiency and reliability.

As for the optimization of BSN applications, one of the most critical issue is to determine the proper conditions that allow to extend the operating life of the wearable devices. Since the radio data transmission and the sensing process (depending on the type of the physical sensor in use) are the most energy-demanding operations, it is reasonable to extend the typical Sensing–Processing–Transmission application pattern with proper autonomic tasks for optimizing such operations depending on some specific conditions and requirements. For instance, as shown in Figure 5.4, the sampling rate can be adapted, at runtime, on the basis of the variability of the resulting output of the processing task. In fact, such that it can be reduced when the data samples do not change so much for a certain period of time, i.e. when the sensor data variability remains below a certain threshold. For the same reason, the radio transmission of such data can be avoided when the application on the coordinator-side does not need to be continuously fed with slightly changing data streams and thus saving energy on the battery-operated sensor nodes by optimizing the most power-hungry operation.

Due to the distinct features of BSNs, like the sensitive nature of data managed (biomedical and personal information), wireless communication, and mobility of sensors, privacy and security represent major concerns for wearable systems to play a significant role in the e-health-care domain. Thus, enhancing the monitoring of physical environment with proper security mechanisms is of prominent importance. The task-based autonomic architecture of SPINE-* does not aim at addressing such issues with specific self-protection mechanisms to contrast external attacks but is intended to provide a proper way to encapsulate such security solutions in reusable components to be directly plugged into the task application when required. As a very simple example, the DataEncryption task of Figure 5.5, which usually demands a lot of

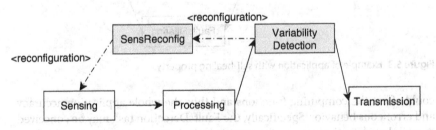

Figure 5.4 Example of application with self-optimization property.

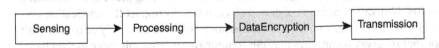

Figure 5.5 Example of application with self-protection property.

computation, can be activated during outdoor activities or when in public untrusted environments but disabled when running at home and secure locations, so as to adapt the application to the execution context.

5.6 Autonomic Physical Activity Recognition

In the following, the benefits of SPINE-* in a real BSN application are presented. Specifically, the existing physical activity recognition described in Refs. [26, 27] is considered as testbed application, which has been turned into an equivalent autonomic version. The whole system consists of a desktop application running on the coordinator and responsible for classifying postures and movements through a k-NN-based classifier to be applied on pre-elaborated data gathered from waist-worn and thigh-worn sensors, both equipped with a 3-axis accelerometer. In particular, the node-side applications consist of (i) sensing the accelerometer sensors, (ii) computing the *mean*, *max*, and *min* features over specific accelerometer axes (also called channels), and (iii) merging and transmitting the results to the coordinator. In Figure 5.6, the two node-side applications, designed through the task-based programming abstraction approach and with no autonomic tasks, are depicted.

Even though such an implementation provides a core functionality for the system to work, it does not include some important features that could be fundamental in case of unexpected conditions. In particular, it is completely unaware of the quality of the data streams coming from the accelerometers and, as a consequence, the activity recognition could provide incorrect detection results. Hereafter, we show how the addition of an autonomic plane, and specifically the integration of self-healing tasks, is beneficial in case of corrupted data. In this regard, the impact of sensed data faults on the activity recognition accuracy is first reported. Then, the improved system fault tolerance and reliability is shown by adopting a proper self-healing layer able to detect and possibly recover such data faults at runtime.

The evaluation approach that we consider consists in carrying out a testbed on the specific predefined sequence of activities shown in Figure 5.7, starting from the "Standing Still" state, and with each state roughly lasting 30 s.

The accelerometer sampling time has been set to 25 ms, whereas the features of the processing tasks (see Figure 5.6) are computed over 40 sampled raw data, every new 20 acquired samples (shift). The k-NN-based classifier has the parameter K set to 1, whereas the Manhattan distance has been adopted due to its excellent performance on well-separated classes (i.e. the states of Figure 5.7). According to such setting and assuming lack of faults in the raw data streams,

(a)

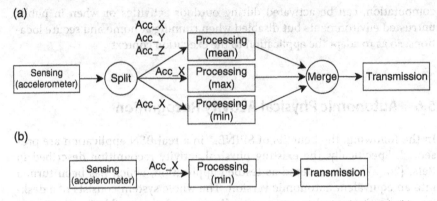

(b)

Figure 5.6 The task-based applications on the waist node (a) and on the thigh node (b).

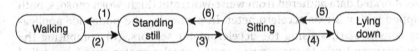

Figure 5.7 The tested activities' sequence.

the classification over the whole activities' transition pattern of Figure 5.7 has obtained an accuracy of 99.75%.

In order to evaluate the impact of erroneous sensor readings on the classification accuracy, we have carried out several tests by considering the original raw data streams and altering them with artificial injected faults before feeding the processing layers of the applications of Figure 5.6. As data faults injection, we consider the models identified in Ref. [22], and in the following, we focus on *short faults*. Such faults consist in irregularities disseminated over a data stream and are modeled as random spikes with parameters P and C, where P is the percentage of raw data affected by spikes and C is the intensity factor, which means that the value of a spike is determined by multiplying the original value of a sensor sample by the C factor.

The results of how the classification accuracy significantly degrades due to short faults are reported in Table 5.1. As a first example, by considering faults affecting all the axes of both accelerometers (on waist and thigh) involved in the preprocessing, the accuracy conspicuously drops to slightly more than 50% with just 5% of the raw data samples affected by spikes.

Also, when considering only one single channel per time affected by faults, the results of Table 5.2 clearly show that the recognition accuracy is more influenced by the quality of the data stream coming from the sensor worn on the thigh, rather than the ones from the waist node.

Hereafter, we show how the introduction of a self-healing plane to the node-side applications of the activity recognition system is capable of improving the

Table 5.1 Activity recognition accuracy affected by short faults over all channels and $C = 3$.

Affected channel	P (%)	Accuracy (%)
All (in both waist and thigh sensors)	1	79.90
All (in both waist and thigh sensors)	5	55.09
All (in both waist and thigh sensors)	10	51.86
All (in both waist and thigh sensors)	25	48.14
All (in both waist and thigh sensors)	50	46.65

Table 5.2 Activity recognition accuracy affected by short faults over a specific channel and $C = 3$.

Affected channel	P (%)	Accuracy (%)
Axis X – Waist sensor	1	98.25
Axis Y – Waist sensor	1	99.75
Axis Z – Waist sensor	1	99.75
Axis X – Thigh sensor	1	81.63
Axis X – Waist sensor	5	96.26
Axis Y – Waist sensor	5	99.75
Axis Z – Waist sensor	5	99.75
Axis X – Thigh sensor	5	44.91

system accuracy by detecting and recovering *short* faults. The enhanced autonomic version of the application in Figure 5.6a is shown in Figure 5.8. In a similar way, a self-healing plane has also been applied on the accelerometer raw data streams of the application running on the thigh node.

With specific reference to the short faults, the underlying approach of the detection and recovery functionalities of the two autonomic tasks is analyzing the variability in the accelerometer data streams. In particular, such streams are split into consecutive data windows, each containing W sensor samples, over which the mean and the standard deviation sd are computed. Then, every single sample in the data window is compared to the standard deviation and, in case its value is much greater than sd, the sample is marked as fault. Specifically, a task parameter T is adopted to determine the threshold value $thr = T \cdot sd$, against which the comparison is performed. If no corrupted data is detected, samples are directly forwarded to the Split task, otherwise the FaultsFiltering will be considered for the subsequent recovering phase. With regards to short

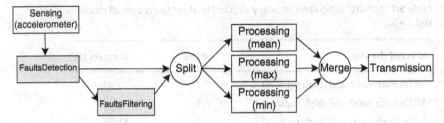

Figure 5.8 The autonomic application running on the waist node.

Table 5.3 Accuracy improvements over all channels and $C=3$.

Affected channel	P (%)	Accuracy (without autonomic plane) (%)	Accuracy (with recovery) (%)
All (in both waist and thigh sensors)	1	79.90	99.75
All (in both waist and thigh sensors)	5	55.09	99.75
All (in both waist and thigh sensors)	10	51.86	98.51
All (in both waist and thigh sensors)	25	48.14	59.55
All (in both waist and thigh sensors)	50	46.65	47.64

faults, the adopted recovering method consists in replacing a corrupted sensor reading with the previous sampled data. Although it appears as a very simple approach, it actually shows its effectiveness in canceling the negative effects of such outliers in the accelerometer streams and thus preventing inaccurate processing leading to low recognition accuracy. A comparison of classification accuracies without and with the self-healing autonomic plane is reported in Table 5.3. In particular, window $W=40$ and threshold parameter $T=3$ have been adopted in our experiments.

With a frequency of short faults within 10%, the recovery operation demonstrates to guarantee very highly accurate outcomes. Conversely, increasingly lower improvements are obtained in the other cases. This is because when a data stream has very recurrent faults, it would be impossible to establish if a specific value is part of a correct data sequence or is a result of a failed sensing operation.

5.7 Summary

Incorporating fault tolerance, adaptability, and reliability into BSNs is a challenging task. In this regard, the autonomic computing is an effective paradigm whose self-* properties are able to fulfill such complex requirements. After having introduced background concepts on the autonomic paradigm, this chapter has presented an architecture for rapid prototyping of BSN applications with autonomic characteristics, SPINE-*. It extends the SPINE2 programming framework by means of an autonomic plane, a way for separating out the provision of self-* properties from the BSN application logic. Then, we have considered a human activity recognition application as a test case by first analyzing how its effectiveness can be heavily affected by data faults in the sensor readings. Finally, we have shown how a self-healing layer (capable of detecting and possibly recovering such faults at runtime) can improve the recognition accuracy, thus improving the quality of the application.

References

1 Horn, P. (2001). Autonomic Computing: IBM's Perspective on the State of Information Technology. *Tech. Rep.*, IBM T.J. Watson Labs, New York.

2 Samaan, N. and Karmouch, A. (2009). Towards autonomic network management: an analysis of current and future research directions. *IEEE Communications Surveys Tutorials* 11 (3): 22–36.

3 Agoulmine, N. (2010). *Autonomic Network Management Principles: From Concepts to Applications*. Ed. Academic Press.

4 The BISON Project Website. http://www.cs.unibo.it/bison (accessed 11 June 2017).

5 The ANA Project Website. www.ana-project.org (accessed 5 June 2017).

6 The Haggle Project Website. http://ica1www.epfl.ch/haggle (accessed 10 June 2017).

7 The CASCADAS Project Website. http://acetoolkit.sourceforge.net/cascadas (accessed 12 June 2017).

8 The EFIPSANS Project Website. http://secan-lab.uni.lu/efipsans-web (accessed 7 June 2017).

9 The Autonomic Internet Project Website. http://www.autoi.ics.ece.upatras.gr/autoi (accessed 8 June 2017).

10 Ruiz, L.B., Nogueira, J.M., and Loureiro, A.A.F. (2003). MANNA: amanagement architecture for wireless sensor networks. *IEEE Communications Magazine* 41 (2): 116–125.

11 Song, H., Kim, D., Lee, K., and Sung, J. (2005). UPnP-based sensor network management architecture. *Second International Conference on Mobile Computing and Ubiquitous Networking (ICMU 2005)*, Osaka, Japan (13–15 April 2005).

12 Lee, W.L., Datta, A., and Cardell-Oliver, R. (2006). WinMS: wireless sensor network-management system, an adaptive policy-based management for wireless sensor networks, *Tech. Rep.*

13 Bourdenas, T. and Sloman, M. (2010). Starfish: policy driven self-management in wireless sensor networks. *Proceedings of the 2010 ICSE Workshop on*

Software Engineering for Adaptive and Self-Managing Systems, ser. SEAMS'10, Cape Town, South Africa (3–4 May 2010), pp. 75–83. New York: ACM.

14 Paradis, L. and Han, Q. (2007). A survey of fault management in wireless sensor networks. *Journal of Network and Systems Management* 15: 171–190.

15 Boonma, P. and Suzuki, J. (2007). Bisnet: a biologically-inspired middleware architecture for self-managing wireless sensor networks. *Computer Networks* 51 (16): 4599–4616, (1) Innovations in Web Communications Infrastructure; (2) Middleware Challenges for Next Generation Networks and Services.

16 Yu, M., Mokhtar, H., and Merabti, M. (2008). Self-managed fault management in wireless sensor networks. *Proceedings of the 2008 the Second International Conference on Mobile Ubiquitous Computing, Systems, Services and Technologies, ser. UBICOMM'08,* Valencia, Spain (29 September–4 October), pp. 13–18. Washington, DC: IEEE Computer Society.

17 Lee, M.-H. and Choi, Y.-H. (2008). Fault detection of wireless sensor networks. *Computer Communications* 31: 3469–3475.

18 Turau, V. and Weyer, C. (2009). Fault tolerance in wireless sensor networks through self-stabilisation. *International Journal of Communication Networks and Distributed Systems* 2: 78–98.

19 Choi, J.-Y., Yim, S.-J., Huh, Y.J., and Choi, Y.-H. (2009). An adaptive fault detection scheme for wireless sensor networks. *Proceedings of the 8th WSEAS International Conference on Software Engineering, Parallel and Distributed Systems,* Cambridge, UK (21–23 February 2009), pp. 106–110. Stevens Point, WI: World Scientific and Engineering Academy and Society (WSEAS).

20 Jiang, P. (2009). A new method for node fault detection in wireless sensor networks. *Sensors* 9 (2): 1282–1294.

21 Oh, H., Doh, I., and Chae, K. (2009). A fault management and monitoring mechanism for secure medical sensor network. *International Journal of Computer Science and Applications* 6: 43–56.

22 Bourdenas, T. and Sloman, M. (2009). Towards self-healing in wireless sensor networks. *Proceedings of the 2009 Sixth International Workshop on Wearable and Implantable Body Sensor Networks, ser. BSN'09,* Berkeley, CA (3–5 June 2009), pp. 15–20. Washington, DC: IEEE Computer Society.

23 Asim, M., Mokhtar, H., and Merabti, M. (2010). A self-managing fault management mechanism for wireless sensor networks. *International Journal of Wireless Mobile Networks* 2 (4): 14.

24 Ji, S., Yuan, S.-F., Ma, T.-H., and Tan, C. (2010). Distributed fault detection for wireless sensor based on weighted average. *Proceedings of the 2010 Second International Conference on Networks Security, Wireless Communications and Trusted Computing – Volume 01, ser. NSWCTC'10,* Wuhan, Hubei, China (24–25 April 2010), pp. 57–60. Washington, DC: IEEE Computer Society.

25 Galzarano, S., Fortino, G., and Liotta, A. (2012). Embedded self-healing layer for detecting and recovering sensor faults in body sensor networks.

Proceedings of the 2012 IEEE International Conference on Systems, Man, and Cybernetics (SMC), Seoul, Korea (14–17 October 2012), pp. 2377–2382.

26 Bellifemine, F., Fortino, G., Giannantonio, R. et al. (2011). SPINE: a domain-specific framework for rapid prototyping of WBSN applications. *Software: Practice and Experience* 41: 237–265.

27 Gravina, R., Guerrieri, A., Fortino, G. et al. (2008). Development of body sensor network applications using SPINE. *IEEE International Conference on Systems, Man and Cybernetics, 2008. SMC 2008*, Singapore (12–15 October 2008), pp. 2810–2815.

6

Agent-Oriented Body Sensor Networks

6.1 Introduction

Many computing paradigms have been to date exploited to support modeling and implementation of wireless sensor networks (WSNs) and, more specifically, of body sensor networks (BSNs). As widely discussed in Chapter 2, different kinds of paradigms, from low level to high level, can be used to develop WSN-based systems. Among such paradigms, the most notable ones are event-driven programming [1], data-based models [2], service-oriented programming [3], macro-programming [4], state-based programming [5], and agent-oriented programming [6]. This chapter proposes the agent-oriented paradigm for the modeling and implementation of BSNs. After introducing background concepts on the agent-computing paradigm and, specifically, on software agents in the WSN context, the chapter discusses motivations and challenges on the exploitation of agents for BSNs and provides a description of the related state-of-the-art. We then present agent-based modeling and implementation of BSNs. A case study is finally proposed that uses two well-known agent-oriented platforms (JADE and MAPS) to develop an agent-based real-time human activity recognition system.

6.2 Background

6.2.1 Agent-Oriented Computing and Wireless Sensor Networks

Software agents are defined as being networked software entities or programs that can perform specific (even complex) tasks for a user and having a degree of intelligence that allows them to carry out parts of their tasks/activities autonomously and to interact with their environment in a useful manner. The features of software agents perfectly fit those of the WSNs and their sensor components [7, 8]; in fact, they mainly include [9]:

Wearable Computing: From Modeling to Implementation of Wearable Systems Based on Body Sensor Networks, First Edition. Giancarlo Fortino, Raffaele Gravina, and Stefano Galzarano.

- *Autonomy*: agents (or sensor nodes) should be able to perform the majority of their problem-solving tasks without the direct intervention of humans, and they should have a degree of control over their own actions and their own internal state.
- *Social ability*: agents (or sensor nodes) should be able to interact, when they deem appropriate, with other software agents (or sensor nodes) and humans in order to complete their own problem solving and to help others with their activities where and when appropriate.
- *Responsiveness*: agents (or sensor nodes) should perceive their environment, in which they are situated and which may be the physical world, a user, a collection of agents (or other sensors), the Internet, etc., and respond in a timely fashion to changes which occur in it.
- *Proactiveness*: agents (or sensor nodes) should not simply act in response to their environment, but they should be able to exhibit opportunistic, goal-directed behavior and take the initiative where and when appropriate.

An interesting taxonomy about WSNs and their relationships with multiagent systems (MAS) can be found in Ref. [8]. In particular, the major motivation of using agents over such networks is that many WSN properties are shared with and can be actually supported by agents and MAS: *physical distribution, resource boundedness, information uncertainty, large-scale, decentralized control*, and *adaptiveness*. Moreover, as sensors in a WSN must typically coordinate their actions to achieve system-wide goals, coordination among dynamic entities (or agents) is one of the main features of MAS. In the following, the aforementioned common properties are discussed:

- *Physical distribution* implies that sensors are situated in an environment from which they can receive stimuli and act accordingly, also through control actions aiming at changing their environment. *Situatedness* is in fact a main property of an agent, and several well-known agent architectures were defined to support such an important property.
- *Boundedness* of resources (computing power, communication, and energy) is a typical property both of sensor nodes as single units and of the WSN as a whole. Agents and related infrastructures can support such limitation through intelligent resource-aware, single, and cooperative behaviors.
- *Information uncertainty* is typical in large-scale WSNs in which both the status of the network and the data gathered to observe the monitored/controlled phenomena could be incomplete. In this case, intelligent (mobile) agents could recover inconsistent states and data through cooperation and mobility.
- *Large-scale* is a property of WSNs either sparsely deployed on a wide area or densely deployed on a restricted area. Agents in MAS usually cooperate in a decentralized way through highly scalable interaction protocols and/or time- and space-decoupled coordination infrastructures.

- *Centralized control* is not feasible in large-scale WSNs as nodes can have intermittent connections and also can suddenly disappear due to energy lack. Thus, *decentralized control* should be exploited. The multiagent approach is usually based on control decentralization transferred either to multiple agents dynamically elected among the available set of agents or to the whole ensemble of agents coordinating as peers.
- *Adaptiveness* is the main shared property between sensors and agents. An agent is by definition adaptive in the environment in which it is situated. Thus, modeling the sensor activity as an agent or a MAS and, consequently, the whole WSN as a MAS could facilitate the implementation of the adaptiveness property.

6.2.2 Mobile Agent Platform for Sun SPOT (MAPS)

MAPS [10–12] is a Java-based framework purposely developed on Sun SPOT technology [13] for enabling agent-oriented programming of WSN applications. MAPS has been developed according to the following requirements:

- *Component-based lightweight agent server architecture* to avoid heavy concurrency by exploiting cooperative concurrency.
- *Lightweight agent architecture* to efficiently execute and migrate agents.
- *Minimal core services* involving agent migration, naming, communication, activity timing, and access to sensor node resources, i.e. sensors, actuators, flash memory, switches, and batteries.
- *Plug-in-based architecture* on the basis of which any service can be defined in terms of one or more dynamically installable components implemented as single or cooperating (mobile) agent(s).
- *Java language* for programming mobile agents.

The architecture of MAPS, shown in Figure 6.1, is based on components that interact through (high level or internal) events and provide a set of services to (mobile) agents including message transmission, agent creation, agent cloning, agent migration, timer handling, and easy access to the sensor node resources.

The main components of the MAPS architecture are described as follows:

- *Mobile Agent* (MA) is the basic high-level component defined by the user for developing agent-based applications.
- *Mobile Agent Execution Engine* (MAEE) controls the execution of MAs by means of an event-based scheduler enabling cooperative concurrency. MAEE also interacts with the other *service-provider* components (see Figure 6.1) to fulfill service requests (e.g. message transmission, sensor reading, and timer setting) issued by MAs.
- *Mobile Agent Migration Manager* (MAMM) supports agents' migration through the Isolate (de)hibernation feature provided by the Sun SPOT

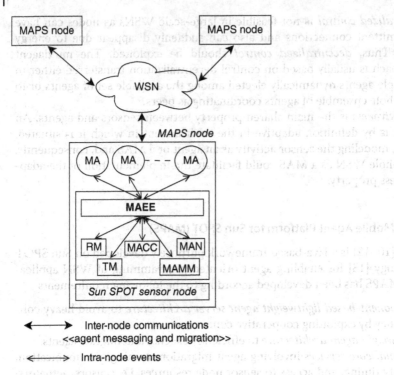

Figure 6.1 Architecture of MAPS.

environment [13]. Such feature involves a data collection and execution state, whereas the agent code should already be at the destination node. This is a limitation of the Sun SPOTs, which do not support dynamic class loading and code migration.

- *Mobile Agent Communication Channel* (MACC) enables interagent communications based on asynchronous messages (unicast or broadcast) supported by the radiogram protocol.
- *Mobile Agent Naming* (MAN) provides agent naming based on proxies for supporting MAMM and MACC in their operations. MAN also manages the (dynamic) list of the neighbor sensor nodes that are updated through a beaconing mechanism based on broadcast messages.
- *Timer Manager* (TM) manages the timer service for timing MA operations.
- *Resource Manager* (RM) manages access to the resources of the Sun SPOT node: sensors (3-axial accelerometer, temperature, and light), switches, LEDs, batteries, and flash memory.

The MAPS Mobile Agent model is depicted in Figure 6.2. Specifically, the dynamic behavior of MA is modeled as a multiplane state machine (MPSM).

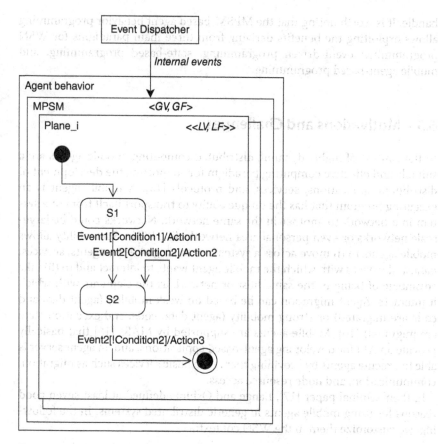

Figure 6.2 Agent behavior model of MAPS.

The GV block represents the global variables, namely, the data inside an MA, whereas the GF is a set of global supporting functions. Each plane may represent the behavior of the MA in a specific role, thus enabling role-based programming [14], and is composed of local variables (LVs), local functions (LFs), and an ECA-based automaton (ECAA). This automaton is composed of states and mutually exclusive transitions among states. Transitions are labeled by Event–Condition–Action (E[C]/A) rules, where E is the event name, [C] is a Boolean expression (or guard) based on global and local variables, and A is an atomic action. A transition is triggered when E is received and C is true. When a triggered transition is fired, A is first atomically executed and then the state transition is completed. MAs interact through events that are asynchronously delivered by the MAEE and dispatched, through the Event Dispatcher component, to one or more planes according to the events that the planes are able to handle. It is worth noting that the MPSM-based agent behavior programming allows exploiting the benefits deriving from three main paradigms for WSN programming: event-driven programming, state-based programming, and mobile agent-based programming.

6.3 Motivations and Challenges

In the context of highly dynamic distributed computing, mobile agents are a suitable and effective computing paradigm for supporting the development of distributed applications, services, and protocols [15]. A mobile agent is an executing program that has the unique ability to transport itself from one system in a network to another in the same network. Networks could be large-scale networks or even personal area networks like BSNs. Such ability allows mobile agents to (i) move across a system containing objects, agents, services, data, and devices with which the mobile agent wants to interact and to (ii) take advantage of being in the same host or network as the elements with which it interacts. Agent migration can be based on weak mobility (agent data and code are migrated) or strong mobility (agent data, code, and execution state are migrated) [16]. Mobile agents are supported by MASs [16] that basically provide an API for developing agent-based applications, and an agent server is able to execute agents by providing them with basic services such as migration, communication, and node resource access.

In their seminal paper [17], Lange and Oshima defined at least seven good reasons for using mobile agents in generic distributed systems. In the following, we customize them in the WSN context:

1) *Network load reduction*: mobile agents are able to access remote resources, as well as communicate with any remote entity, by directly moving to their physical locations and interacting with them locally to save bandwidth resources. For instance, a mobile agent that incorporates data-processing algorithm/s can move to a sensor node (e.g. a wearable sensor node), perform the needed operations on the sensed data, and transmit the results to a sink node. This is more desirable, rather than executing a periodic transmission of raw sensed data from the sensor node to the sink node and the data processing on the latter.

2) *Network latency overcoming*: an agent provided with proper control logic may move to a sensor/actuator node to locally perform the required control tasks. This overcomes the network latency that will not affect the real-time control operations also in case of lack of network connectivity with the base station.

3) *Protocol encapsulation*: if a specific routing protocol supporting multi-hop paths should be deployed in a given zone of a WSN, a set of cooperating

mobile agents encapsulating this protocol can be dynamically created and distributed into the proper sensor nodes without any regard for standardization matter. Also in case of protocol upgrading, a new set of mobile agents can easily replace the old one at runtime.

4) *Asynchronous and autonomous execution*: these distinctive properties of mobile agents are very important in dynamic environments like WSNs where connections may not be stable and network topology may change rapidly. A mobile agent, upon a request, can autonomously travel across the network to gather required information "node by node" or to carry out the programmed tasks and, finally, can asynchronously report the results to the requester.

5) *Dynamic adaptation*: mobile agents can perceive their execution environment and react autonomously to changes. This behavioral dynamic adaptation is well suited for operating on long-running systems like WSNs where environment conditions are very likely to change over time.

6) *Orientation to heterogeneity*: mobile agents can act as wrappers among systems based on different hardware and software. This ability can fit well the need for integrating heterogeneous WSNs supporting different sensor platforms or connecting WSNs to other networks (like IP-based networks). An agent may be able to translate requests coming from a system into suitable ones for another different system.

7) *Robustness and fault tolerance*: the ability of mobile agents to dynamically react to unfavorable situations and events (e.g. low battery level) can lead to better robust and fault-tolerant distributed systems; e.g. the reaction to the low battery level event can trigger a migration of all executing agents to an equivalent sensor node to continue their activity without interruption.

6.4 State-of-the-Art: Description and Comparison

Although many MASs [18] were developed for conventional distributed platforms, a very few agent frameworks for WSNs have been to date proposed and concretely implemented. In the following, we first describe Agilla and actorNet, the most significant available research prototypes based on TinyOS [19], and then, we overview AFME and MAPS, which are the most representative ones based on the Java language.

Agilla [6] is an agent-based middleware developed on TinyOS and supporting multiple agents on each node. It provides two fundamental resources on each node:

- The *tuplespace*, which represents a shared memory space where structured data (tuples) can be stored and retrieved, allowing agents to exchange information through spatial and temporal decoupling. A tuplespace can be also accessed remotely.
- The *neighbor list*, which contains the address of all one-hop nodes needed when an agent has to migrate.

Agilla agents can migrate carrying their code and state, but they cannot carry their tuples locally stored on a tuplespace. Packets used for node communication (e.g. for agent migration/cloning and remote tuple accessing) are very small to minimize losses, whereas retransmission techniques are also adopted.

ActorNet [20] is an agent-based platform specifically designed for Mica2/ TinyOS sensor nodes. To overcome the difficulties of code migration and interoperability due to the strict coupling between applications and sensor node architectures, actorNet exposes services like virtual memory, context switching, and multitasking. Due to these features, actorNet effectively supports agent programming by providing a uniform computing environment for all agents, regardless of hardware or operating system differences. The actor-Net language used for high-level agent programming has syntax and semantics similar to those of Scheme with proper instruction extension.

Both Agilla and actorNet are designed for TinyOS that relies on the nesC language.

The Java language, through which Sun SPOT [13] and Sentilla JCreate [21] sensors can be programmed, due to its object-oriented features, could provide more flexibility and extendibility for an effective implementation of agent-based platforms. Currently, the main available Java-based mobile agent platforms for WSNs are MAPS [11] and AFME [22].

The AFME framework [22], a lightweight version of the Agent Factory framework purposely designed for wireless pervasive systems and implemented in J2ME, is also available on Sun SPOT and is used for exemplifying agent communication and migration in WSNs. AFME is strongly based on the Belief–Desire–Intention (BDI) paradigm, in which intentional agents follow a sense–deliberate–act cycle. In AFME, agents are defined through a mixed declarative and imperative programming model. The declarative Agent Factory Agent Programming Language (AFAPL), based on a logical formalism of beliefs and commitments, is used to encode an agent's behavior by specifying rules that define the conditions under which commitments are adopted. The imperative Java code is instead used to encode perceptors and actuators. However, AFME was not specifically designed for WSNs and, particularly, for Java Sun SPOT.

MAPS, the Java-based agent platform overviewed in Section 6.2.2, is conversely specifically designed for WSNs and currently uses the release 4.0 (Blue) of the Sun SPOT library to provide advanced functionality of communication, migration, timing, sensing/actuation, and flash memory storage. MAPS allows developers to program agent-based applications in Java according to the rules of the MAPS framework, and thus no translator and/or interpreter need to be developed and no new language has to be learnt as in the case of Agilla, ActorNet, and AFME. MAPS was also ported on the Sentilla JCreate sensor platform and renamed TinyMAPS [21].

In Table 6.1 a comparison among the aforementioned agent platforms is reported.

Table 6.1 Comparison among agent-oriented platforms (Agilla, ActorNet, AFME, and MAPS) for WSNs.

	Agilla	ActorNet	AFME	MAPS
Agent migration availability	Yes	Yes	Yes	Yes
Concurrent agents	Yes	Yes	Yes	Yes
Agent communications	Tuple-based	Asynchronous messages	Asynchronous messages	Asynchronous messages
Agent programming language	Proprietary ISA	Scheme-like	Declarative + Java	Java
Agent model	Assembly-like	Functional	BDI	Finite state machine
Intentional agents availability	No	No	Yes	No
WSN-supported platforms	Mica2, MicaZ, TelosB	Mica2	Sun SPOT	Sun SPOT, Sentilla JCreate

6.5 Agent-Based Modeling and Implementation of BSNs

As widely discussed in Chapter 1, a BSN is basically composed of a coordinator node or base station and one or more wearable sensor nodes connected with a 1-hop wireless connection with the coordinator. According to the agent-oriented approach, each component of a system is agentified; therefore, both the BSN coordinator and the BSN sensor nodes are modeled as agents. A BSN system as a whole constitutes a MAS that is basically structured as a master/slave system (see Figure 6.3a), where the coordinator is the master agent and the sensor nodes are the slave agents. The slave agents can only interact with the coordinator agent. A variant of the basic architecture (see Figure 6.3b) is a mix of Master/Slave (M/S) and peer-to-peer (P2P): the coordinator agent can interact with all slave agents and the slave agents can interact with each other. Both basic M/S and advanced M/S + P2P can be used to structure a single BSN. To model collaborative/interacting BSNs (see Chapter 7), the Super Peer model (see Figure 6.3c) can be exploited: coordinator agents are super peers and can interact with each other, whereas sensor nodes belonging to the same BSN can only interact with each other and with their coordinator agent.

Agent-based implementation of BSN systems should be based on real agent platforms [23] supporting the programming of both the coordinator agent and the application agents and the sensor agents. Specifically, we propose JADE [24] to implement the application and coordinator agents and MAPS [11] to implement the sensor agents. Thus, agent programming follows the rules of JADE and MAPS. Moreover, the application development of agent-based applications is also supported by an agent-oriented software engineering methodology [25], which usually covers the phases of requirement analysis, design, implementation, and deployment. In the next section, a case study is proposed to exemplify the agent-based engineering approach for BSN applications.

6.6 Engineering Agent-Based BSN Applications: A Case Study

In order to show the effectiveness of agent-based platforms to support programming of BSN applications, in Ref. [26] a MAPS-based agent-oriented signal-processing in-node environment specialized for real-time human activity monitoring has been proposed. In particular, the system is able to recognize postures (e.g. lying down, sitting, and standing still) and movements (e.g. walking) of assisted livings. The architecture of the developed agent-based system, shown in Figure 6.4, is organized into three types of agents:

- The *Application-level Agent* (running on a PC or a handheld device) that embeds the application logic, implemented with Java and JADE [24].

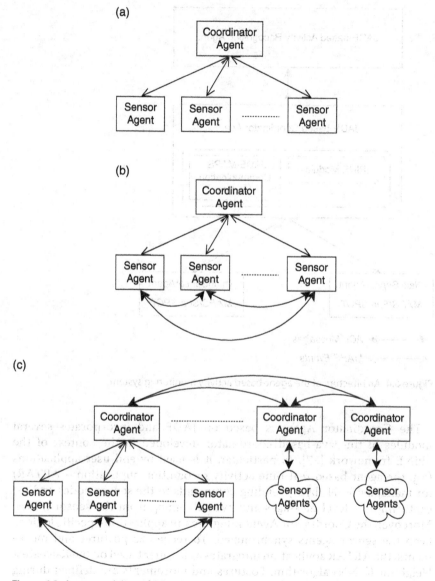

Figure 6.3 Agent modeling of BSNs. (a) Master/slave model, (b) Master/slave+peer-to-peer model, and (c) Super peer model.

- The *Coordinator Agent* (running on a PC or a handheld device), implemented with Java and JADE.
- The *Sensor Agent* (running on the wearable sensor nodes), programmed with MAPS [11].

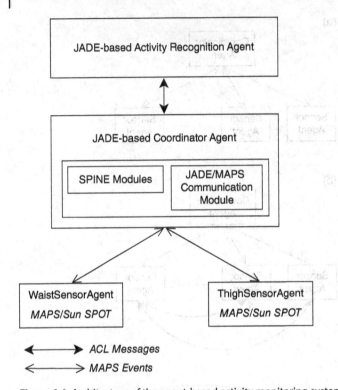

Figure 6.4 Architecture of the agent-based activity monitoring system.

The Coordinator Agent is based on JADE and incorporates several modules of the Java-based coordinator developed in the context of the SPINE framework [27]. In particular, it is used by end-user applications (e.g. the agent-based real-time activity recognition application – ARTAR) for managing BSNs by (i) sending commands to the sensor nodes and (ii) capturing low-level messages and events coming from the sensor nodes. Moreover, the Coordinator Agent integrates an application-specific logic to keep the sensor agents synchronized. To recognize postures and movements, the ARTAR application integrates a classifier based on the K-Nearest Neighbor (k-NN) algorithm. Postures and movements are defined during the training phase. ARTAR and the Coordinator Agent interact through JADE ACL messages.

While the ARTAR and the Coordinator Agent are based on JADE, the two sensor agents are based on MAPS. Thus, a communication adaptation module between JADE and MAPS was developed to allow communication interoperability. The two sensor nodes are, respectively, positioned on the waist and the thigh of the monitored-assisted living. Specifically, two sensor agents are defined: *WaistSensorAgent* and *ThighSensorAgent*. Their behaviors are modeled through a 1-plane MPSM (see Section 6.2.2) by executing the following stepwise cycle:

1) *Accelerometer Data Sensing*: the 3-axial accelerometer sensor collects raw accelerometer data (<*Acc_X*, *Acc_Y*, and *Acc_Z*>) according to a given sampling time.
2) *Feature Computation*: specific features are computed on the collected raw accelerometer data. Features are calculated as follows: (i) Mean on all accelerometer axes for the WaistSensorAgent, (ii) Max and Min on the X accelerometer axis for the WaistSensorAgent, and (iii) Min on the X accelerometer axis for the ThighSensorAgent.
3) *Feature Merging and Transmission*: the computed features are merged into a single message and transmitted to the Coordinator Agent.
4) *Go to 1*.

Figure 6.5 also shows how such elaboration cycle is actually programmed using the MAPS finite state machine.

In Ref. [26], the entire BSN system has been analyzed in depth by considering the following two aspects:

• The performance evaluation of the timing granularity degree of the sensing activity at the sensor node and the synchronization degree or skew of the activities of the two sensor agents.
• The recognition accuracy that shows how well the human postures and movements are recognized by the overall agent system.

On the basis of the obtained performance results, it can be stated that MAPS shows its great suitability for supporting efficient BSN applications, thus demonstrating that the agent approach is not only effective during the design and implementation phases of a BSN application but also during its execution. Furthermore, the recognition accuracies are good and encouraging if compared with other works in the literature that use more than two sensors on the human body to recognize activities [28]. Finally, with reference to the programming effectiveness of MAPS, the MAPS programming model based on the finite state machine offers a very straightforward and intuitive tool for supporting BSN application development.

6.7 Summary

This chapter has provided an overview of the use of the agent-oriented paradigm to model and implement BSN systems. We have first introduced the motivations and challenges for this exploitation. Then, we have introduced

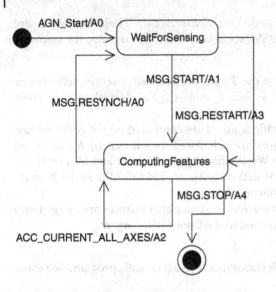

AGN_Start: event to start the agent behavior plane
MSG.START: event to start the sensing activity
MSG.RESTART: event to restart the sensing activity
MSG.RESYNCH: event to resynchronize the agent timing
MSG.STOP: event to stop the agent activity
ACC_CURRENT_ALL_AXES: event including raw sensed data

A0: Initialize the local variables of the plane
A1: Initialize the buffers to store raw sensed data
  Create the timer for sensor sampling
  Launch the sensing activity
A2: Fill the buffer with the raw sensed data
  Compute the features after N sensor samplings and
  Transmit the features to the coordinator agent
  Create the timer for sensor sampling
  Launch the sensing activity
A3: Disable the sensing timer
  Initialize the local variables of the plane
  Execute action A1
A4: Disable the sensing timer

Figure 6.5 Finite state machine of the sensor agents: WaistSensorAgent and ThighSensorAgent.

MAPS for WSN-based system development. Furthermore, related work and a qualitative comparison among the most diffused (mobile) agent platforms for WSNs have been discussed. Finally, the chapter has focused on agent-oriented BSN application development based on MAPS; specifically, a MAPS-based human activity recognition BSN system has been described.

References

1 Gay, D., Levis, P., von Behren, R. et al. (2003). The nesC language: a holistic approach to networked embedded systems. *Proceedings of the ACM SIGPLAN 2003 Conference on Programming Language Design and Implementation*, San Diego, CA (9–11 June 2003).

2 Madden, S.R., Franklin, M.J., Hellerstein, J.M., and Hong, W. (2005). TinyDB: an acquisitional query processing system for sensor networks. *ACM Transactions on Database Systems (TODS)* 30 (1): 122–173.

3 Marin, C. and Desertot, M. (2005). Sensor bean: a component platform for sensor-based services. *Proceedings of the 3rd International Workshop on Middleware for Pervasive and Ad-Hoc Computing, MPAC'05*, Grenoble, France (28 November–2 December 2005), pp. 1–8. ACM.

4 Gummadi, R., Gnawali, O., and Govindan, R. (2005). Macroprogramming wireless sensor networks using Kairos. *Proceedings of the International Conference on Distributed Computing in Sensor Systems (DCOSS)*, Fortaleza, Brazil (10–12 June 2015).

5 Kasten, O. and Römer, K. (2005). Beyond event handlers: programming wireless sensors with attributed state machines. *Proceedings of the 4th International Symposium on Information Processing in Sensor Networks*, Los Angeles, CA (24–27 April 2005).

6 Fok, C.-L., Roman, G.-C., and Lu, C. (2009). Agilla: a mobile agent middleware for sensor networks. *ACM Transactions on Autonomous and Adaptive Systems* 4 (3): 1–26.

7 Rogers, A., Corkill, D., and Jennings, N.R. (2009). Agent technologies for sensor networks. *IEEE Intelligent Systems* 24: 13–17.

8 Vinyals, M., Rodriguez-Aguilar, J.A., and Cerquides, J. (2010). A survey on sensor networks from a multiagent perspective. *The Computer Journal* 54 (3): 455–470.

9 Wooldridge, M.J. and Jennings, N.R. (1995). Intelligent agents: theory and practice. *The Knowledge Engineering Review* 10 (2): 115–152.

10 Aiello, F., Fortino, G., Gravina, R., and Guerrieri, A. (2009). MAPS: a mobile agent platform for Java Sun SPOTs. *Proceedings of the 3rd International Workshop on Agent Technology for Sensor Networks (ATSN-09)*, jointly held with the *8th International Joint Conference on Autonomous Agents and Multiagent Systems (AAMAS-09)*, Budapest, Hungary (12 May 2009).

11 Aiello, F., Fortino, G., Gravina, R., and Guerrieri, A. (2011). A Java-based agent platform for programming wireless sensor networks. *The Computer Journal* 54 (3): 439–454.

12 MAPS – Mobile Agent Platform for Sun SPOT. Documentation and software. http://maps.deis.unical.it (accessed 23 August 2015).

13 Sun SPOT. Documentation and code. www.sunspotdev.org (accessed 14 June 2017).

14 Zhu, H. and Alkins, R. (2006). Towards role-based programming. *Proceedings of CSCW'06*, Banff, Alberta (4–8 November 2006).

15 Yoneki, E. and Bacon, J. (2005). A survey of Wireless Sensor Network technologies: research trends and middleware's role. *Tech. Rep. UCAM-CL-TR-646*, University of Cambridge.

16 Karnik, N.M. and Tripathi, A.R. (1998). Design issues in mobile-agent programming systems. *IEEE Concurrency* 6: 52–61.

17 Lange, D.B. and Oshima, M. (1999). Seven good reasons for mobile agents. *Communications of the ACM* 42 (3): 88–90.

18 Fortino, G., Garro, A., and Russo, W. (2008). Achieving mobile agent systems interoperability through software layering. *Information & Software Technology* 50 (4): 322–341.

19 TinyOS. Documentation and software. www.tinyos.net (accessed 9 June 2017).

20 Kwon, Y., Sundresh, S., Mechitov, K., and Agha, G. (2006). ActorNet: an actor platform for wireless sensor networks. *Proceedings of the 5th International Joint Conference on Autonomous Agents and Multiagent Systems (AAMAS)*, Hakodate, Japan (28 April 2006), pp. 1297–1300.

21 Aiello, F., Fortino, G., Galzarano, S., and Vittorioso, A. (2012). TinyMAPS: a lightweight Java-based mobile agent system for wireless sensor networks. In *Fifth International Symposium on Intelligent Distributed Computing (IDC2011)* (5–7 October), Delft, the Netherlands. In *Intelligent Distributed Computing V, Studies in Computational Intelligence*, 2012, Vol. 382/2012, pp. 161–170. doi: 10.1007/978-3-642-24013-3_16.

22 Muldoon, C., O'Hare, G.M.P., O'Grady, M.J., and Tynan, R. (2008). Agent migration and communication in WSNs. *Proceedings of the 9th International Conference on Parallel and Distributed Computing, Applications and Technologies*, Dunedin, New Zealand (1–4 December 2008).

23 Luck, M., McBurney, P., and Preist, C. (2004). A manifesto for agent technology: towards next generation computing. *Autonomous Agents and Multi-Agent Systems* 9 (3): 203–252.

24 Bellifemine, F., Poggi, A., and Rimassa, G. (2001). Developing multi agent systems with a FIPA-compliant agent framework. *Software Practice and Experience* 31: 103–128.

25 Fortino, G. and Russo, W. (2012). ELDAMeth: an agent-oriented methodology for simulation-based prototyping of distributed agent systems. *Information & Software Technology* 54 (6): 608–624.

26 Aiello, F., Bellifemine, F., Fortino, G. et al. (2011). An agent-based signal processing in-node environment for real-time human activity monitoring based on wireless body sensor networks. *Journal of Engineering Applications of Artificial Intelligence* 24: 1147–1161.

27 Bellifemine, F., Fortino, G., Giannantonio, R. et al. (2011). SPINE: a domain-specific framework for rapid prototyping of WBSN applications. *Software: Practice and Experience* 41 (3): 237–265.

28 Maurer, U., Smailagic, A., Siewiorek, D.P., and Deisher, M. (2006). Activity recognition and monitoring using multiple sensors on different body positions. *Proceedings of the International Workshop on Wearable and Implantable Body Sensor Networks (BSN'06)*, Cambridge, MA (3–5 April 2006), pp. 113–116. IEEE Computer Society.

7

Collaborative Body Sensor Networks

7.1 Introduction

The importance of wearable systems in facilitating and empowering many human-centered domains has been already widely proved and discussed. However, despite their potential, the current BSN-based systems are mostly used for applications focusing on the monitoring of a single individual. Also, the current BSN frameworks aim at providing effective programming supports for easily and efficiently developing applications for remote, real-time monitoring of assisted livings over network based on a multisensors/single-coordinator configuration. Since more and more applications in several domains (health care, entertainment, social interaction, sport, and emergency among others) demand different and more complex BSN-based architectures, the paradigm centered on a single individual monitoring is not sufficient anymore to meet these new applications' requirements.

Thus, the need for new multi-BSN infrastructures, henceforth indicated as *Collaborative Body Sensor Networks* (*CBSNs*), is compelling to foster novel applications based on the collaborative approach of groups of individuals, where single BSNs have to cooperate with each other to properly monitor and recognize group activities in order to fulfill a common goal.

In this chapter, a reference architecture for CBSN applications, thus enabling interactions among single BSNs, is presented. Moreover, a new programming framework, *Collaborative SPINE* (*C-SPINE*) [1,2], specifically designed to fully implement the proposed CBSN architecture, is also described. Proposed as an enhancement of the SPINE framework [3,4] (see Chapter 3), it provides specific data communication, multisensor data fusion, collaborative processing, and joint data analysis capabilities to facilitate the development of novel smart wearable systems for the current and future cyber-physical pervasive computing environments.

Wearable Computing: From Modeling to Implementation of Wearable Systems Based on Body Sensor Networks, First Edition. Giancarlo Fortino, Raffaele Gravina, and Stefano Galzarano.
© 2018 John Wiley & Sons, Inc. Published 2018 by John Wiley & Son, Inc.

7.2 Background

Most of the current applications using wearable systems rely on BSN infrastructures constituted of a collection of sensor nodes wirelessly connected to a single coordinator device (the base station – BS), which usually makes the individual's information locally or remotely available.

However, today's complex application scenarios require more dynamic and flexible interacting components and thus new types of BSNs need to be defined in order to offer further capabilities. In the following, the possible kinds of BSN infrastructures are introduced. They are categorized on the basis of the "logical interconnections" among the main communicating BSN components, i.e. the individuals wearing the sensor nodes and the coordinators/BSs (depicted as smartphones), with no assumption about the actual underlying physical network topologies. As depicted in Figure 7.1, we have the following logical BSN infrastructures:

a) *Single Body–Single Base station (SBSB)* (Figure 7.1a): the wearable devices of a single individual communicate with a single BS. This is the most common configuration for the current available body-monitoring applications aimed at acquiring, processing, and storing (locally or remotely) the biomedical signals of individuals.

b) *Single Body–Multiple BSs (SBMBs)* (Figure 7.1b): such a configuration enables communications between a single BSN with multiple BSs. A typical scenario could be in the home automation context, where an individual may interact with BSs located in different places of the environment.

c) *Multiple Bodies–Single BS (MBSB)* (Figure 7.1c): multiple BSNs can be coordinated by a single BS, which allows for indirect interaction among different individuals. An example is in the gaming context, where a device (smart-TV or a game console) enables an augmented social experience in a group of people wearing sensors.

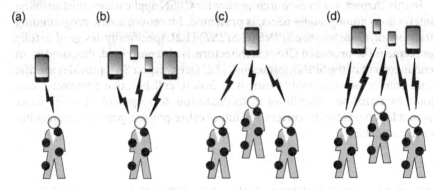

(a)   (b)   (c)   (d)

Figure 7.1 BSN infrastructures based on the logical communications among individuals and base stations.

d) *Multiple Bodies–Multiple BSs* (*MBMBs*) (Figure 7.1d): multiple BSNs can interchangeably and dynamically communicate with multiple BSs. This configuration is needed in more complex scenarios (e.g. during large-scale disaster), where the emergency intervention by a team of rescuers requires a more efficient and automatic coordination and a better delivery of information related to the victims' conditions.

7.3 Motivations and Challenges

Based on the characteristics of the different BSN infrastructures, none of the BSN-specific solutions or programming frameworks developed and proposed so far (discussed in Chapter 2) is specifically designed to straightforwardly support multi-BSN configuration. In fact, most of them are designed around the basic needs to perform multisensor data fusion [5] in single-BSN contexts and are usually implemented by following a three-layer architecture:

1) *Sensing layer*: this module provides sensor sample acquisition and signal data gathering functionality from the on-body sensors. Besides extracting raw data, it usually also computes basic feature extraction functions such as min, max, mean, variance, etc.
2) *Analysis layer*: starting from the set of extracted features, this layer is in charge of selecting and joining the most significant ones by further providing some decision algorithms, like a classifier.
3) *Dissemination layer*: high-level information from the analysis layer is delivered to some user-applications, which can either be locally (i.e. on the coordinator/ BS device) or remotely executed.

Such a three-layer architecture lacks fundamental capabilities to support inter-BSN communication and collaborative, distributed processing functionality, which are needed to successfully support the *Multiple Bodies–Multiple BS* configuration.

Hence, the novel reference architecture for CBSNs proposed in this chapter has been purposely conceived to fully adhere to all the possible BSN configurations. Such a general architecture has been later exploited as a guideline for implementing a supporting framework aimed at facilitating the development of collaborative BSN applications. In particular, the need for a CBSN infrastructure can be better motivated by the fact that it is capable of easily enabling new services allowing single BSNs to interact with each other (not addressed in the other BSN configurations):

• *Client/Server services*: a pair of BSNs can interact in a standard client/server communication paradigm, where a server BSN (e.g. the monitored individual) provides services to let the client BSN issue (i) a continuous monitoring request or (ii) a single data request. In the former, the server BSN continuously pushes information to the client, whereas the latter works as a more typical single-reply-upon-request model.

- *Broadcast services*: BSNs can broadcast (push) information without being queried about (i) the individuals' worn sensors or (ii) alarm/events triggered by the individuals' conditions (e.g. a critical status like a fall or a heartbreak).
- *Collaborative services*: aimed at performing specific tasks upon direct interactions between BSNs and based on a peer-to-peer model to exchange information. They usually detect and recognize group activities and relevant events based on the implicit or explicit multiuser interactions.

Despite their benefits, such services pose further challenges in implementing a CBSN application framework, which needs to successfully fulfill new specific requirements:

- *Inter-BSN communication*: all the aforementioned types of service models require reliable and robust inter-BSN communication mechanisms to be implemented.
- *BSN proximity detection*: providing proper proximity detection protocols is of fundamental importance for managing the neighbor CBSNs.
- *Discovery of BSN services*: complementary to the BSN proximity detection system, a CBSN should rely on dynamic but well-specified (possibly standard) service discovery methods.
- *Selection and activation of BSN services*: in a similar way, selecting and activating services need common mechanisms to be implemented according to a specific protocol.
- *Collaborative multisensor data fusion*: specific distributed algorithms for group activities' classification/detection represent major tasks (and challenges) in the CBSN context.

7.4 State-of-the-Art

As already discussed, most of the proposed BSN solutions or programming frameworks developed so far are not conceived to straightforwardly support multi-BSN infrastructures, since they are basically designed around a multisensor data fusion approach and implemented by following the three-layer architecture presented in the previous section.

In Ref. [6], a postures and activities recognition system by fusing data from multiple two-axial accelerometers is proposed. The angular velocity, and the horizontal and vertical accelerations of sensors placed on different locations of the human body are estimated by Kalman Filters (KFs) and the resulting flexion angles of body parts, which are real-time indicators of limbs and torso position, are in turn processed (using statistical, temporal, and spectral features) and identified, based on banks of trained Hidden Markov Models (HMMs), and fused together to infer whole body posture or activity.

The authors in Refs. [4,7] propose a human activity recognition system using two 3-axial accelerometers placed on the right thigh and on the waist. Specific features are extracted from data collected on a fixed-length periodic window basis: max, min, average and total energy on all axes of the accelerometer on the waist, and max on the x axis of the accelerometer of the thigh sensor. Such features are then merged and classified by using a k-NN-based decision tree to identify different activities, like standing still, sitting, lying down, walking, and fall events along with the extent of the fall.

In Ref. [8], different human postures (sitting, squatting, standing still, and lying down) are detected by using a multisensor data fusion method that relies on the D-S evidence theory. Empirical evidence is extracted from 3-axial accelerometers placed on calf, thigh, arm, and waist so that ranges of the gravity acceleration can be defined for each axis and related to each activity of interest. The theory of evidence is exploited to first define basic trust functions, which are then combined, on the basis of real-time observations, to generate more accurate functions to recognize the postures.

In Ref. [9], a novel multiobjective Bayesian Framework for Feature Selection (BFFS) and a method for searching optimal solutions are proposed. It can be used in BSN systems for reducing the number of relevant features by eliminating the redundant ones and thus identifying the sensors that do not considerably influence the decision process. Moreover, a contextual multisensor data fusion method, based on model learning and inferencing algorithms, is proposed to recognize individual's activities.

The authors in Ref. [10] propose self-healing methods to detect data faults from sensors. Specifically, it is shown how the accuracy of a BSN system for human activity recognition is affected by different types of faulty data. Some filtering methods are then proposed to improve sensor data quality and enhance the recognition accuracy.

In Ref. [11], the authors aim at improving classification robustness against sensor failures by proposing a formulation of a latent structure influence model capable of capturing the correlation among (including noisy/faulty) different sensing processes. A BSN system able to recognize eight locations, six speaking/non-speaking states, six postures, and eight activities is considered as a case study.

7.5 A Reference Architecture for Collaborative BSNs

The proposed reference architecture for supporting a CBSN infrastructure can be described under two different perspectives, the networking perspective and the functional one:

- *Network Architecture*, which shows the communications among BSNs in terms of basic and the application-specific interaction protocols.
- *Functional Architecture*, which defines the types and activities of the main functional blocks in charge of managing the general system and executing some specific tasks depending on the actual applications.

As depicted in Figure 7.2, the CBSN Network Architecture consists of different sets of wearable sensors (the WSs) and base-stations (BSs). In the picture, we assume that every CBSN is controlled by a BS, which manages the sensor nodes through an application-level intra-BSN communication, usually implemented over a single-hop protocol based on a physical star topology. The interaction between a pair of BSs is made through an Inter-BSN protocol. In the case of absence of a BS, the set of WSs constituting the CBSN may be directly accessed by other BSs through the Intra-BSN protocol (IBP).

In the following, the list of functions offered by the IBP is provided:

- *Service discovery*, for retrieving the available services (processing, sensing, and actuating) for each of the WS composing the CBSN.
- *Service configuration*, for setting the parameters of the discovered WS services.

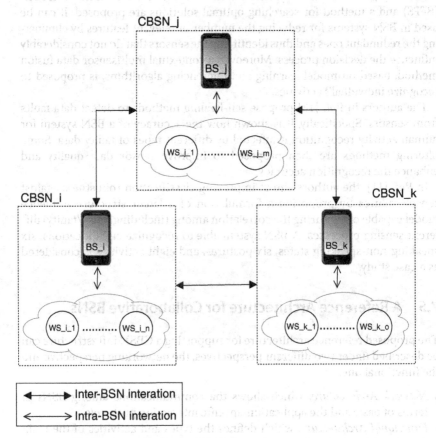

Figure 7.2 The reference CBSN Network Architecture.

- *Service control,* used to manage the operations on the WS services, i.e. activate/deactivate, monitor, and control a configured service.
- *Data transmission,* for exchanging raw and/or processed data between the BS and the WSs of the same CBSN.

The inter-BSN interaction is enabled by some application-specific protocols, which support the collaboration among high-level applications and services running on each CBSN. In addition, in order to provide some basic common operations, a set of protocols should be defined: *Proximity Detection Protocol* (PDP), *Service Discovery Protocol* (SDP), and *Service Selection and Activation Protocol* (SSAP). The activity diagram shown in Figure 7.3 represents the flow of these basic inter-BSN common operations. In particular, the PDP is intended to detect and manage other CBSNs in the neighboring location by means of a beaconing approach. When a CBSN is detected, the SDP is used for sharing and managing the list of the available services that each CBSN can provide to the others: at first, a service description request is broadcast, and upon its reception, a reply containing services information is communicated. The SSAP is in charge of actually controlling and managing calls to one or more selected specific services required by a certain application. Once activated and executed, such collaborative services interact by exchanging some service-specific messages.

Figure 7.4 depicts the CBSN *Functional Architecture,* which includes the following components at BS side (some of them have been already previously described):

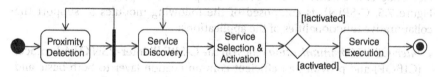

Figure 7.3 Activity diagram of basic CBSN operations.

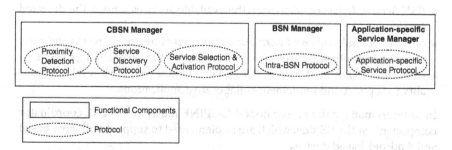

Figure 7.4 CBSN Functional Architecture components.

- *CBSN Manager* manages the first three basic operations, by using the PD, SD, and SSA protocols. In particular, a service can be either automatically activated upon a service discovery or activated on demand by the CBSN owner. The former approach usually relies on some mutual knowledge relationship among owners.
- *BSN Manager* handles the WSs belonging to the CBSN through the IBP.
- *Application-specific Service Manager* manages and executes the application-level services through the Application-specific Service Protocol (ASP, see the next point).
- *Application-specific Service Protocol (ASP)* implements the communication mechanisms for allowing the interaction among services related to the final applications.
- *Proximity Detection Protocol, Service Discovery Protocol, Service Selection and Activation Protocol* implement the mechanisms for CBSN proximity detection and service discovery, selection, and activation.
- *Intra-BSN Protocol (IBP)* is for coordinating the interaction between the WSs and the BS.

7.6 C-SPINE: A CBSN Architecture

A full-fledged CBSN middleware, named Collaborative SPINE (C-SPINE), has been developed as an implementation of the reference architecture described in Section 7.5. It includes the sensor-side and BS-side components of SPINE (see Chapter 3) besides CBSN-specific components. In particular, as shown in Figure 7.5, C-SPINE is composed of the following modules to support the collaborative functionalities of the applications:

- *Inter-CBSN Communication* relies on the C-SPINE Inter-BSN OTA Protocol (CIBOP) and provides an efficient communication layer to both basic and application-specific services and protocols.
- *BSN Proximity Detection* implements the procedure for detecting neighbor CBSNs.
- *BSN Service Discovery* discovers the available services among the detected CBSNs.
- *BSN Service Selection and Activation* implements the mechanisms and rules to select and activate discovered services among the surrounding CBSNs.
- *Application-Specific Protocols and Services* are a set of higher level functionality to support and implement collaborative applications.

In order to manage the sensor nodes, C-SPINE reuses the SPINE coordinator components at the BS side, which are implemented to support both Java-based and Android-based devices:

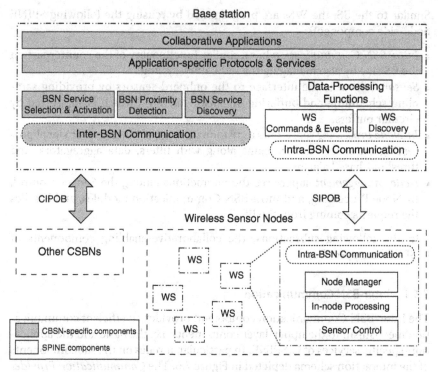

Figure 7.5 C-SPINE Functional Architecture components.

- *Intra-BSN Communication* handles the message transmission and reception according to the SPINE Intra-BSN OTA Protocol (SIBOP). It abstracts away from the specific WS platform-related communication protocol by using the proper radio module. It currently provides radio support for TinyOS motes and Sun SPOT devices.

- *WS Commands and Events* offers developers the interface to coordinate the BSN by allowing to activate sensing and processing functions on the nodes as well as handling BSN events (e.g. new discovered nodes, alarms, and user data messages) and forwarding them to the registered application-level modules.

- *WS Discovery* manages the discovery functionality of WS nodes.

- *Data-Processing Functions* module provides developers the interface for a set of signal processing, feature extraction, pattern recognition, and data classification functions in order to facilitate the development of new applications. The module also provides an adaptation with the WEKA Data Mining toolkit [12].

Similar to the BS, the WSs are programmed by reusing the following SPINE node-side components:

- *Intra-BSN Communication* has a similar functionality of the counterpart on the BS side by also managing the radio duty-cycling.
- *Sensor Control* is the interface to the onboard sensors by providing sampling scheduling and buffering of sensor readings, which is supported by circular buffers.
- *In-Node Processing* represents a customizable set of functions for signal processing on sensor data streams along with filters, data aggregators, and threshold-based alarms.
- *Node Management* supervises the interactions among the Sensor Control, In-Node Processing, and Intra-BSN Communication modules, and handles the requests coming from the BS.

In the following subsections, the collaborative-enabling components of C-SPINE are described.

7.6.1 Inter-BSN Communication

The Inter-BSN Communication component provides an efficient communication mechanism to the upper-layer components, i.e. the basic and the application-specific services of C-SPINE. In particular, it relies on the subcomponents of the interaction schema depicted in Figure 7.6. The *Communication Provider (CP)* is in charge of managing the exchange of messages among CBSNs and thus provides a set of methods for configuring a CBSN to receive and send messages of specific types. Each different type of message requires a specific

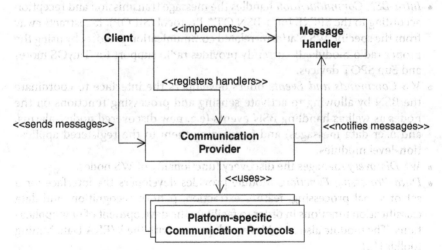

Figure 7.6 Inter-BSN component interaction.

Message Handler (MH) component in order to be correctly processed, and every MH needs to be registered with the CP so as to be notified of new incoming messages.

A specific CIBOP protocol, which depends on the upper-level services and application requirements, is defined and implemented according to such design schema. In particular, the following steps have to be accomplished to correctly define a new interaction protocol (IP):

1) Defining a new univocal message type identifying the new protocol.
2) Creating a set of IP-specific messages, all belonging to the same previously defined message type.
3) Implementing a MH, linked to the new message type, to handle and interpret the new set of messages.
4) Registering the MH with the CP.

Since the Inter-BSN Communication provides an abstract mechanism to support higher level communication layers, it includes a set of adapters in order to use real platform-specific lower level communication protocols. In particular, C-SPINE currently supports both the Bluetooth and the IEEE 802.15.4 protocols, which are dynamically chosen depending on the actual physical platform of the BS.

7.6.2 BSN Proximity Detection

Based on a beaconing mechanism, the Proximity Detection component is designed around network-driven adaptation approaches for controlling the beacon rate and managing the neighbor cache (each CBSN handles a table containing information about its neighbor CBSNs). The beacon rate, defined in terms of frequency f_{hello}, depends on the network conditions and specifically on the *turnover rate* (r_t) value, which is computed as:

$$r_t = \frac{N_{ndn}}{N_{nc}}$$

where N_{ndn} is the number of new discovered CBSNs, whereas N_{nc} is the total amount of the currently cached CBSNs. In case the r_t value is less than a specific threshold r_{opt}, the f_{hello} value is reduced (the beacon interval time is increased by Δt, which is usually set equal to 500 ms), as a result of few changes that occurred in the proximity. Conversely, if r_t is greater than the threshold, the beacon interval is incremented by Δt.

As for the cache of neighbors, the information regarding each neighboring CBSN *cb* is stored in the history table as a tuple having the following structure:

$$< beacon\_time(cb),\ T_1(cb),\ T_2(cb),\ Wait(cb) >$$

where *beacon_time(cb)* is the timestamp of the last beacon received from *cb*, $T_1(cb)$ and $T_2(cb)$ are the reception intervals of the last two beacons, and *Wait(cb)* is the amount of time after which the neighboring CBSN *cb* is removed from the cached table if a new beacon is not received. In particular, *Wait(cb)* is updated as follows:

$$Wait(cb) = \begin{cases} k * T_1(cb) & \text{if } T_1(cb) = T_2(cb) \\ T_1(cb) + \dfrac{T_1(cb)}{|T_1(cb) - T_2(cb)|} & \text{if } |T_1(cb) - T_2(cb)| \geq 1 \\ T_1(cb) + T_1(cb) * \left(|T_1(cb) - T_2(cb)|\right) & \text{if } 0 < |T_1(cb) - T_2(cb)| < 1 \end{cases}$$

7.6.3 BSN Service Discovery

This component is in charge of discovering the set of available services between pairs of interacting CBSNs. Specifically, C-SPINE provides two different service discovery mechanisms: *on-demand* and *advertisement-driven*. The former approach allows to directly query one or more neighbor CBSNs, among the ones detected by the Proximity Detection component, for obtaining the list of the provided services. The *advertisement-driven* service discovery relies on the advertisement messages, containing the list of offered services, which are periodically broadcast along with the beaconing messages.

7.6.4 BSN Service Selection and Activation

The Service Selection and Activation component allows pairs of CBSNs to mutually make use of their respective discovered services in order to accomplish specific collaborative tasks required by the running applications. In particular, selecting and activating a service is performed through the specification of well-defined rules; it additionally depends on the mutual acquaintance relation between interacting CBSNs and possibly on some contextual information. A service-selecting rule is defined by the following tuple:

$$< ID_S, R_A [, CTX] >$$

where:

- ID_S is a numeric identification code specifying a certain service.
- $R_A \subseteq ID^n{}_{CBSN}$ (with $n \geq 2$) represents the relation among two or more CBSNs on the basis of their mutual acquaintance. The following annotation are examples of relations: $<IDx, *>$ identifies public services, $<IDx, IDy>$ indicates a service that is enabled only between a pair of CBSNs, whereas $<ID_1, ID_2, ..., ID_n>$ enables a group of CBSNs to use the service. If the interacting CBSN identifier is a component of such relation, R_A holds.

- CTX, which is an optional attribute, specifies a logical (e.g. *walking*) or physical context (e.g. *home* or *hospital*) in which the interaction takes place. If the interacting CBSN has this attribute, CXT holds.

A rule holds if and only if both R_A and CXT (if any) hold. Thus, the service can be selected and activated. Moreover, according to the defined rules, selection and activation of services can be manually (i.e. driven by the user) or automatically configured.

7.7 Summary

Despite the BSN technology is of fundamental importance in enabling and facilitating the development of many human-centered applications, most of the current systems have been designed and implemented for simply being employed in the monitoring of single individuals. However, new application scenarios are demanding different BSN-based architectures requiring a novel paradigm based on multi-BSN cooperation in order to properly accomplish more complex collaborative tasks. Thus, this chapter has focused on the motivations and requirements for which the stand-alone BSN approach is not suitable anymore and a novel reference architecture for Collaborative BSNs (CBSNs) has been described. Also, a new programming framework, called C-SPINE and evolved from the SPINE basic structure, has been presented as a real implementation of the aforementioned CBSN reference architecture.

References

1 Fortino, G., Galzarano, S., Gravina, R., and Li, W. (2014). A framework for collaborative computing and multi-sensor data fusion in body sensor networks. *Information Fusion* 22: 50–70.
2 Augimeri, A., Fortino, G., Galzarano, S., and Gravina, R. (2011). Collaborative body sensor networks. *Proceedings of the 2011 IEEE International Conference on Systems, Man, and Cybernetics (SMC)*, Anchorage, AK (9–12 October), pp. 3427–3432.
3 Fortino, G., Giannantonio, R., Gravina, R. et al. (2013). Enabling effective programming and flexible management of efficient body sensor network applications. *IEEE Transactions on Human-Machine Systems* 43 (1): 115–133.
4 Bellifemine, F., Fortino, G., Giannantonio, R. et al. (2011). SPINE: a domain-specific framework for rapid prototyping of WBSN applications. *Software: Practice & Experience* 41 (3): 237–265. doi: 10.1002/spe.998.
5 Khaleghi, B., Khamis, A., Karray, F.O., and Razavi, S.N. (2013). Multisensor data fusion: a review of the state-of-the-art. *Information Fusion* 14 (1): 28–44. http://dx.doi.org/10.1016/j.inffus.2011.08.001.

6 Dong, L., Wu, J., and Chen, X. (2007). Real-time physical activity monitoring by data fusion in body sensor networks. *2007 10th International Conference on Information Fusion*, Quebec, Canada (9–12 July 2007), pp. 1–7. doi: 10.1109/ICIF.2007.4408176.

7 Gravina, R., Guerrieri, A., Fortino, G. et al. (2008). Development of body sensor network applications using SPINE. *IEEE International Conference on Systems, Man and Cybernetics (SMC)*, Singapore (12–15 October), pp. 2810–2815, doi: 10.1109/ICSMC.2008.4811722.

8 Li, W., Bao, J., Fu, X. et al. (2012). Human postures recognition based on D–S Evidence theory and multi-sensor data fusion. *Proceedings of the 12th IEEE/ ACM International Symposium on Cluster, Cloud and Grid Computing, ccGRID 2012, IEEE Computer Society*, Ottawa, Canada (13–16 May), pp. 912–917. doi: 10.1109/CCGrid.2012.144.

9 Thiemjarus, S. (2007). A framework for contextual data fusion in body sensor networks. PhD thesis. Imperial College London.

10 Bourdenas, T. and Sloman, M. (2009). Towards self-healing in wireless sensor networks. *Proceedings of the 2009 Sixth International Workshop on Wearable and Implantable Body Sensor Networks, BSN'09*, Berkeley, CA (3–5 June 2009), pp. 15–20. Washington, DC: IEEE Computer Society. doi: 10.1109/ BSN.2009.14.

11 Dong, W. and Pentland, A. (2006). Multi-sensor data fusion using the influence model. *Proceedings of the International Workshop on Wearable and Implantable Body Sensor Networks, BSN'06*, Cambridge, MA (3–5 April 2006), pp. 72–75. Washington, DC: IEEE Computer Society. doi: 10.1109/ BSN.2006.41.

12 Witten, I.H., Frank, E., and Hall, M.A. (2011). *Data Mining: Practical Machine Learning Tools and Techniques*. Boston, MA: Morgan Kaufmann Publishers.

8

Integration of Body Sensor Networks and Building Networks

8.1 Introduction

This chapter provides a research- and technical-oriented perspective on the integration of body sensor networks (BSNs) and Building Networks (BNs), which are based on wireless sensor and actuator networks (WSANs). The aim of this integration is twofold: (i) supporting indoor wearable computing based on BSNs through a data collection and provision infrastructure offered by BNs and (ii) seamlessly including data coming from BSNs into WSAN-based infrastructures like BNs. This integration would therefore enable the construction of human-centered smart environments ranging from smart buildings to fully automated ambient-assisted living contexts. After providing some fundamentals on BNs, and presenting the motivations and challenges related to the BSN/ BN integration, the chapter focuses on the definition of the integration layers according to a networking-based approach. We will then discuss and compare the state-of-the-art about BSN/WSN integration with respect to the defined layers. Finally, the chapter presents an agent-oriented gateway for the integration of BSNs, based on SPINE (see Chapter 3), and BNs, based on the Building Management Framework. Moreover, a set of diverse human-centered smart environments that can be supported through the proposed gateway and, more generally, through BSN/WSN integration is also enumerated.

8.2 Background

8.2.1 Building Sensor Networks and Systems

A wireless sensor network (WSN) [1] is a collection of tiny devices capable of sensing, computation, and wireless communication operating in a certain environment to monitor and control events of interest in a distributed manner and to collectively react to critical situations. WSN applications span various domains such as environmental and building monitoring and surveillance, pollution monitoring, agriculture, health care, home-automation, energy management, earthquake, and eruption monitoring. WSNs applied in the context of buildings are typically referred as building sensor and actuator networks [2], or simply BNs. An example of BN environment is shown in Figure 8.1. BNs aim at satisfying different needs of inhabitants of buildings such as awareness regarding their structural health, control over the building environment, actuation of specific policies in the energy management of buildings, trade-off with respect to energy consumption and people comfort, support for context-aware social and commercial activities, safety, and security. Differently from pure WSNs, in BNs actuators are fundamental components to regulate devices and thus control the building environment. Examples of systems for BNs are described in Refs. [2–7]. In Ref. [8], the authors propose a set of qualitative indicators that can be used for analyzing the above-cited BN systems and notably for developing new ones; indeed, they can also be considered requirements specifically elicited for building management systems based on WSANs:

- *In-node data processing*: executing processing on the nodes in a BN allows to create and send synthetic packets in the network and to reduce the amount of raw data toward the base station, thus decreasing the energy consumption on the nodes (the radio is the most energy-consuming component of the nodes). Moreover, reducing the amount of packets created by the BN nodes allows more nodes to share the same radio channel.

Figure 8.1 An example of the Building Network environment.

- *Multi-hop network protocols*: due to the short radio range that the BN nodes can cover, a framework for the building management has to provide support for multi-hop networks relying, for example, on specific data-centric or hierarchical protocols [9].
- *Fast network (re)configuration*: when BN nodes are already placed, it is too onerous in terms of time, and sometimes difficult, reaching all the nodes to reconfigure them. This means that a framework for BNs has to provide mechanisms to quickly (re)configure BN nodes. This is usually done through optimized configuration packets sent over the air.
- *Support for heterogeneous devices*: the building management can require the use of particular sensor boards available only for particular sensor platforms or different computation power in different nodes of the BN. To provide this flexibility, a framework for the building management should provide multi-platform support for the inclusion of heterogeneous devices.
- *Support for actuators*: managing actuators in a building is fundamental since they allow to remotely control devices in order to apply particular policies to achieve specific building-wide goals such as comfort or energy saving.
- *Abstractions to model the building floor plan*: since BN nodes can be deployed everywhere in a building, it is useful that they are aware of their position. Moreover, a coordinator should have the possibility to program/query nodes on the basis of their physical and logical characteristics (either if the node is in a specific position, such as a room or close to a window, or if it has particular sensors/actuators, such as the temperature sensor). To offer this service, a framework for the building management should provide a set of programming abstractions to model the floor plan of a building. Typically, to support these programming abstractions, nodes in a BN are organized in sets of logical or physical groups that may also partially overlap [2].
- *Decision delocalization*: in a BN, an important feature that reduces the packets toward the base station is the delocalization of some functions. Specific nodes in a building can have the capacity to take some decisions, controlling actuators, or collecting data from their neighbors to make data aggregation. For example, a node can collect temperature data in a room and send to the base station only the mean temperature over all the nodes of that room or a node can decide to switch a radiator on if the temperature in a room goes below a certain threshold.
- *Deploy management through human–computer interface*: a framework for the building management should provide an extensible and user-friendly graphical interface to easily manage the BN. The GUI should allow to effectively (re)configure the BN and visually present the data coming from the network.
- *Multi-base station organization for large-scale BNs*: when the scale of a BN starts to be very large, like in skyscrapers, industrial warehouses, or

multibuilding constructions, the *tree depth* of the network can become very big and, consequently, every packet in the BN should follow too many hops to reach the base station. This results in a big waste of batteries and consequential reduction of the network lifetime. To reduce such a phenomenon, a framework for the BN management should provide instruments to manage large-scale environments. A multi-base station organization of the BN can address this problem. In particular, every base station can have its independence and share with the other base stations only what is needed. Such base stations can, for example, be developed as software agents as it has been done in Ref. [4].

- *Remote management of the BN*: not always a local and centralized management is what BN users require. Often, especially for buildings that are large or with more than one administrator, a remote control of the BN is needed. To provide such functionalities, several approaches can be used. In Ref. [10], for example, a gateway approach was used to allow the remote programming of a BN and decouple the GUI from the BN base station.

8.2.2 Building Management Framework

The Building Management Framework [2, 11] is a domain-specific framework implemented for both WSAN nodes and more capable devices at the coordinator (or base station) side such as PCs, plug computers, smartphones, and PDAs. The BMF allows flexible and efficient distributed sensing and actuation in buildings and in all other contexts in which sensors/actuators can be deployed in environments and on physical objects. BMF provides fast reconfiguration, in-node processing algorithms, multi-hop routing, hw/sw multiplatform support, a building programming abstraction (named dynamic groups) to dynamically model the morphology of buildings and physical spaces, support for actuators, and an extensible application programming interface. The BMF architecture, portrayed in Figure 8.2 consists of two-layered software components at coordinator side and sensor node side. The coordinator and the sensor nodes interact through the application-level BMF communication protocol based on a multi-hop network protocol. Moreover, applications can use a high-level interface (BMF API) to communicate with the coordinator. At the coordinator, the *Request Scheduling* layer provides an API through which requests for programming sensing and actuating operations can be easily created and scheduled. Requests can address single nodes or groups of nodes that can be dynamically created. At the node side, the *Multi-Request Scheduling* layer is able to execute multiple requests sent from the coordinator. Interested readers can find an in-depth description of all BMF components and protocols, along with application examples, in Refs. [2, 11].

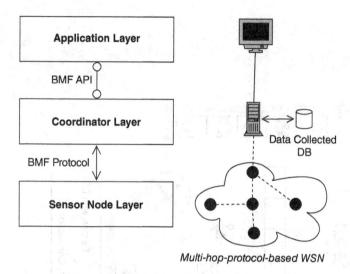

Multi-hop-protocol-based WSN

Figure 8.2 The overall BMF framework architecture.

8.3 Motivations and Challenges

The integration of BNs and BSNs aims at facilitating the development of novel smart environments, namely human-aware smart buildings, effectively supporting people while entering and moving inside (residential, commercial, public, and private) buildings. Figure 8.3 shows a building floor environment embedding wireless sensors and hosting people that wear BSNs.

Main provided services, which can be defined through the BN/BSN integration, could be categorized into *basic* and *advanced*:

- *Basic services*
  - *People identification*, which is fundamental to identify people inside the building.
  - *People localization*, which allows to trace the location of people inside the building.
  - *Information exchange*, which enables the transfer of different kinds of information between people and the smart building. For instance, the smart building could monitor the vital parameters of people for healthcare assistance.
- *Advanced services*
  - *Safety*, which supports people in case of emergency. For example, this service could suggest the safest pathway/s to exit the building in case of a fire alarm.

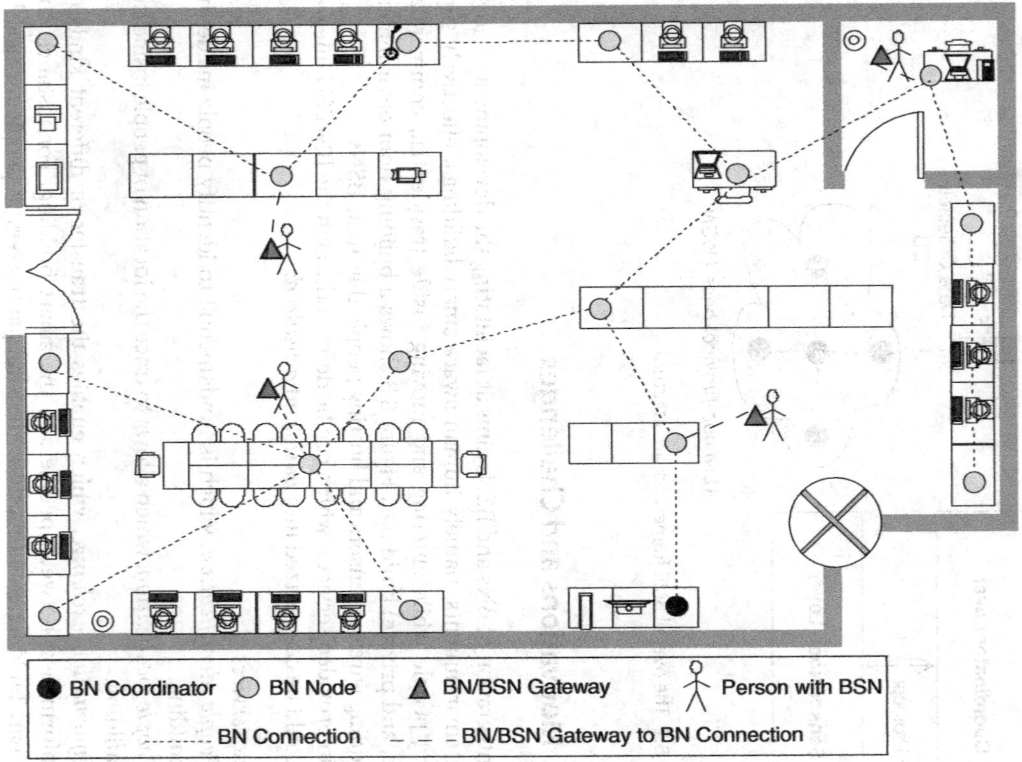

Figure 8.3 BN/BSN integration: a scenario.

Legend:
- ● BN Coordinator
- ○ BN Node
- ▲ BN/BSN Gateway
- ⚲ Person with BSN
- —— BN Connection
- ----- BN/BSN Gateway to BN Connection
- ----- BN Connection

- *Security*, which supports the security of building by monitoring author-ized/unauthorized people and enforcing space access.
- *Context-aware personal support*, which is based on the first three basic services and provides specific services depending on the type of buildings and context in which people are located. For instance, in a commercial building such as a mall, the smart building could send advertisements to people depending on their captured emotions while approaching and visiting shops.

8.4 Integration Layers

Different types of BN/BSN integration can be envisaged at different networking layers (physical, MAC, network, and application) (see Figure 8.4):

- *BN and BSN use the same protocols*: in this case, BN and BSN have to be homogeneous (same physical, MAC, network, and application layers) so that BSN nodes seamlessly become members of the BN.
- *BN and BSN only have different physical layers*: in this case, BN and BSN have to be homogeneous at the MAC, network, and application layers and have to interact through hubs in the network that translate the data between different physical media.
- *BN and BSN have different physical and MAC layers*: in this case, BN and BSN have to be homogeneous at the network and application layers and have to interact through bridges in the network that translate the data between different MAC layers. Moreover, bridges can apply filtering on the MAC addresses of the packets that are not addressed to the subnet they manage.

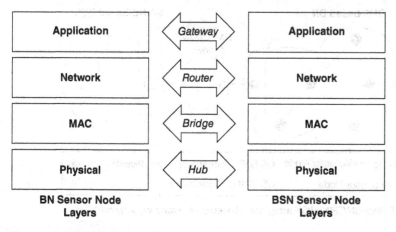

Figure 8.4 BN/BSN integration layers.

- *BN and BSN have different physical, MAC, and network layers*: in this case, BN and BSN have to be homogeneous only at the application layer and have to interact through routers that merge networks running different network protocols (usually BNs use multi-hop network protocols while BSNs use star-topology single-hop protocols). Routers can filter data based on destination addresses.
- *BN and BSN implement different physical, MAC, network, and application layers*: in this case, BSN and BN need to interact through an application gateway. So, BSN and BN are independent and share a node that acts as a gateway between the two different networks. This node knows both the BSN and the BN communication protocols at all layers and will translate data between the networks at the application layer.

Among the discussed integration approaches, we believe that the application gateway is the most suitable and viable one because it allows to use different protocol stacks for BNs and BSNs and also different transmission media. This also allows for a high degree of heterogeneity of the involved devices (coordinators, sensors, and actuators) and avoids interoperability issues at different layers. The most suitable node on which to install the gateway is represented by the BSN coordinator as we can assume that each BSN has a powerful coordinator (smartphone, tablet, and PDA) with (i) a specific node interfacing the BN and actually being a (mobile) node of the BN and (ii) a specific node interfacing with the BSN nodes. A specific gateway-based solution is shown in Figure 8.5, where the application-level gateway interfaces BMF-based BNs with SPINE-based BSNs. Such a solution will be implemented in Section 8.6 through an agent-oriented approach.

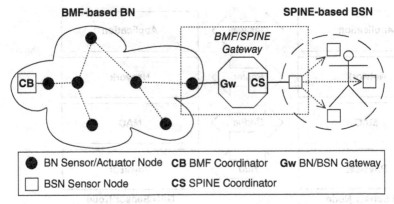

Figure 8.5 BMF-BN/SPINE-BSN integration based on the gateway approach.

8.5   State-of-the-Art: Description and Comparison

The integration of heterogeneous networked systems is an important problem that has been addressed in different research and industrial contexts so far.

In Ref. [12], the authors designed and implemented the NETA Monitoring System, which is based on standard agents standing on different platforms. NETA addressed the problem of integrating autonomous and heterogeneous IT systems that are not correlated, thus allowing for automatic monitoring across systems that would otherwise require manual intervention. These agents report, in an asynchronous fashion, events to a System Engine, which is the core of the NETA Monitoring System. It is in charge of correlating events and managing any trouble for each platform.

Integration of different classes of networks is instead the aim of Buddhikot et al. [13]. The development of the integration approach is based on the introduction of two components in the system: a new network element called IOTA (Integration Of Two Access technologies) gateway deployed in the network and a new client software. In particular, the IOTA gateway cooperating with the client software offers integrated 802.11/3G wireless data services that support seamless inter-technology mobility, Quality of Service (QoS) guarantees, and multiprovider roaming agreements.

In Ref. [14], the authors design and implement a framework that uses mobile agents to ensure information exchange between legacy network management systems. Their aim is the realization of an evolutionary network redesign that preserves the existing infrastructure and saves the operator's existing investments. The framework is based on layered decentralized management architecture and implemented using agents on the network and subnet layers.

In Ref. [15], the authors present a novel agent-based approach to data translation between WSNs and an existing agent-based air condition monitoring system. Their aim is to demonstrate that a multiagent approach combined with wireless sensor networking can be used for a number of air condition monitoring applications. They designed and implemented a sensor network gateway that provides an interface between the JADE FIPA-based multiagent system and the WSN.

In Ref. [16], the authors present the design and implementation of the JADE/ MAPS gateway. It allows integration between two agent platforms, namely JADE which is used for conventional distributed environments, and MAPS (see Section 6.2.2), which is exploited in WSNs. Thus, the gateway enables also the integration of distributed platforms and WSNs. The gateway has been implemented as a JADE agent to provide a communication mechanism between JADE and MAPS agents, thus facilitating bidirectional translation between JADE ACL messages and MAPS events and supporting routing of communication between the two agent platforms.

In Ref. [17], an integrated communications framework for context-aware continuous monitoring with BSNs is proposed. This is the most representative work, along with the one described in the next section, related to the integration of BSNs and WSNs. In particular, the paper proposes a wireless pervasive communication system to support advanced health-care applications. The system is based on an ad-hoc interaction of mobile BSNs with independent WSNs already deployed within the environment to allow a continuous and context-aware health monitoring for assisted livings along their daily-life scenarios. Specifically, the proposal is at the MAC level: a novel MAC layer protocol, namely MD-STAR, is proposed, aiming at improving the capabilities of synchronization/localization in a scenario in which a mobile BSN interacts with fixed WSNs. However, the system is only evaluated through simulation, so no real implementation exists.

8.6 An Agent-Oriented Integration Gateway

The architecture of the gateway solution [18], shown in Figure 8.5, has been developed through an agent-oriented approach based on JADE [19]. In particular, the JADE-based gateway is a multiagent system composed of two interacting JADE agents: the BMFAgent and the SPINEAgent.

The BMFAgent interfaces the BMF network by encapsulating and enhancing the behavior of a BMF node. From the BMF network perspective, the BMFAgent is just a BMF node (see Section 8.2.2) interacting with the BMF coordinator by using the BMF protocols.

The SPINEAgent interfaces the SPINE network by encapsulating the SPINE coordinator (see Chapter 3). From the SPINE network perspective, the SPINEAgent is just a SPINE coordinator interacting with the SPINE nodes through the SPINE protocols.

The class diagram of the agent-based gateway including the BMFAgent and the SPINEAgent is reported in Figure 8.6.

The BMFAgent is composed of the following classes:

- *BMFAgent*, which is the main BMFAgent class, extends the JADE Agent class and keeps track of all the instantiated behaviors.
- *BMFInteraction*, which is the component allowing the interaction with the BMF-based BN, implements the BMF communication protocol [2].
- *BMFBehavior*, which interprets the requests sent from the BN and instantiates new one shot or periodic behaviors, communicates with the SPINEAgent through ACL-based messages to get the list of the available sensors in the SPINE-based system.
- *OneShotBehavior*, which is the behavior that allows managing one-shot requests (either threshold-based or not), interacts with the SPINEAgent through ACL-based messages to receive data from sensors.

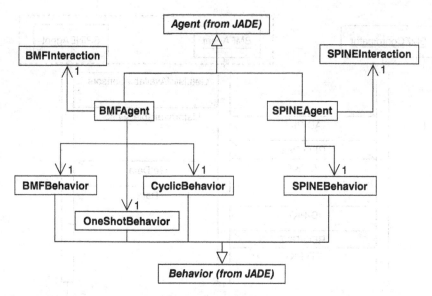

Figure 8.6 Class diagram of the agent-based gateway.

- *CyclicBehavior*, which is the behavior managing Periodic requests (either threshold-based or not), interacts through ACL-based messages with the SPINEAgent to retrieve data from sensors.

The SPINEAgent consists of the following classes:

- *SPINEAgent*, which is the main SPINEAgent class, extends the JADE Agent class.
- *SPINEInteraction*, which is the component allowing the interaction with the SPINE-based BSN system, implements the SPINE communication protocol (see Chapter 3).
- *SPINEBehavior*, which is the component allowing the interaction with the BMFAgent through ACL-based messages, provides the list of available sensors, collects data from sensors, and sends sampled data to the BMFAgent.

The ACL-based interaction between the pair <BMFAgent and SPINEAgent> and the BMF Coordinator is reported in the interaction diagram of Figure 8.7. Specifically, as soon as the gateway is activated, the BMFAgent sends a request to the SPINEAgent for retrieving the list of available sensing services. A sensing service is based on either real or virtual sensors. When the BMFAgent receives the reply from the SPINEAgent, it sends the advertisement message (AD-PKT) to the BMF Coordinator, advertising the available sensing services. Such message is indeed sent periodically. As soon as the BMF Coordinator receives the advertisement message, it includes the agent-based gateway in the

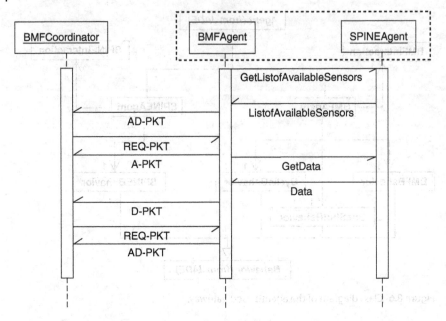

Figure 8.7 Interaction between the agent-based gateway (pair <BMFAgent, SPINEAgent>) and the BMF Coordinator.

BMF network as a real BMF node. From now on, the BMF Coordinator can issue request messages (REQ-PKT) targeting the BMFAgent that, in turn, replies with an acknowledgment message (A-PKT). The BMFAgent is able to interpret three types of sensing requests:

- *One-shot*, which allows to request a single reading of raw or aggregated sensed data from selected sensors.
- *Periodic*, which allows to set up periodic readings of raw or aggregated sensed data from selected sensors.
- *Threshold-based*, which allows to configure a single reading (or periodic readings) of raw or aggregated sensed data from selected sensors when such data are compliant with the defined threshold-based operations (>t, <t, >=t, <=t, =t, in [t1, t2]).

After request interpretation, the BMFAgent creates and adds a JADE OneShotBehavior executing the simple or threshold-based one-shot request or a JADE CyclicBehavior executing the simple or threshold-based periodic request. Such behaviors are able to request data to the SPINEAgent and, according to the request logic, process received data and send data messages (D-PKT) to the BMF Coordinator. If the BMF Coordinator wants to stop any data message from the BMFAgent, a reset message (RS-PKT) can be sent to the BMFAgent that will then start sending AD-PKT to the BMF Coordinator.

Finally, the gateway has a mechanism dealing with mobility [20]: a problem which can arise in this kind of scenarios is that a gateway can be temporarily off-line because it is far from any BN node or because the handoff procedure (i.e. the gateway detaches from one BN node and attaches to another BN node) is not instantaneous. In this case, some data packet from the gateway to the BMF Coordinator can be lost. To overcome this problem, at the low level of the gateway an intelligent buffer has been implemented; it stores the data to be sent to the BMF Coordinator and, once online, sends all the buffered data to the coordinator.

8.7   Application Scenarios

The BN/BSN integration promotes the development of diversified smart environments such as AAL (Ambient Assisted Living) environments [21] and human-centered smart buildings [3]. Physical activity recognition and monitoring is a basic building block that enables both the aforementioned application domains. Indeed, physical activity recognition is one of the fundamental building blocks of many BSN applications [22]. It is often necessary to monitor daily activity levels for wellness applications; it may help identifying abnormal heart rate variations, e.g. by correlating the heart rate variations with the current activity being performed, and it can be even applied in highly interactive computer games, to cite a few scenarios. Smart environments can monitor activities of their inhabitants to better support them for basic and customized services (see Section 8.3). In the following subsection, an in-building human activity monitoring system is designed through the agent-based approach proposed in Section 8.6.

8.7.1   In-Building Physical Activity Monitoring

The proposed in-building human activity monitoring system architecture is shown in Figure 8.8. The overall system consists of the BMF Coordinator, the BMF WSAN network, and the BMF/SPINE agent-based gateway connected to a SPINE-based BSN system. In particular, the SPINE-based BSN system [23, 24] uses only two wireless motion sensor nodes placed on the waist and on the thigh of the assisted living, and a personal smart-phone running an activity recognition application, which is able to detect the following four basic activities: *lying down*, *sitting*, *standing*, and *walking*. This is achieved with or without an individual training phase, and with an overall average accuracy of about 98% [23]. Furthermore, the BSN system may also report the number of steps performed by the subject and detect the event of accidental falls that may potentially lead to dangerous situations (e.g. after a detected fall, the system also recognizes how

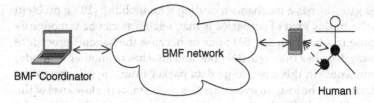

BMF Coordinator

Human i

Figure 8.8 Architecture of the in-building physical activity recognition system.

Table 8.1 Sensing services of the BSN system for human activity recognition

Sensing service	Description	Values
Activity	Activity performed	{"lying down", "sitting", "standing", "walking"}
Step counter	Number of steps walked	Integer
Fall	Person falling	True/false
AccWaist	3-Axial acceleration of the sensor worn on the waist	(AccX, AccY, AccZ)
AccThigh	3-Axial acceleration of the sensor worn on the thigh	(AccX, AccY, AccZ)

long the subject is lying down and, according to a given threshold, it can trigger an alert message). The complete list of the sensing services provided by the BSN system is reported in Table 8.1.

The BN system allows for different monitoring modes of the sensing services that can be easily and dynamically programmed by the BMF Coordinator:

- *Continuous*, which supports continuous acquisition of the sensing service data according to a programmable sampling rate.
- *On-demand*, which allows to query the sensing service when needed.
- *Alert-based*, configures specific thresholds on the sensing service data; when such thresholds are satisfied, a notification is sent from the sensing service.

It is worth noting that the BN system can not only simply monitor the activity of humans in the building but also detect specific transitions (e.g. sit-to-stand) or critical events (e.g. falling) and, on their detection, send out alerts. Such system feature is essential to configure personalized monitoring on the basis of people identity and fulfill specific single and collective objectives.

8.8 Summary

This chapter has proposed the integration of BSNs and BNs, i.e. WSANs for monitoring and automation of buildings. We have first introduced the motivations and challenges for such integration. We have then introduced a layered architecture enabling integration at different network layers. Furthermore, related works and their comparison according to this architecture have been discussed. Then, the chapter has focused on an agent-oriented integration gateway, actually enabling the integration of SPINE-based BSNs and BMF-based WSANs. Finally, a smart environment for physical activity recognition featured by the proposed integration approach has been analyzed.

References

1 Akyildiz, I.F., Su, W., Sankarasubramaniam, Y., and Cayirci, E. (2002). Wireless sensor networks: A survey. *Computer Networks: The International Journal of Computer and Telecommunications Networking* 38 (4): 393–422.

2 Fortino, G., Guerrieri, A., O'Hare, G.M.P., and Ruzzelli, A.G. (2012). A flexible building management framework based on wireless sensor and actuator networks. *Journal of Network and Computer Applications* 35 (6): 1934–1952.

3 Snoonian, D. (2003). Smart buildings. *IEEE Spectrum* 40: 18–23.

4 Fortino, G. and Guerrieri, A. (2012). Decentralized management of building indoors through embedded software agents. *Computer Science and Information Systems* 9 (3): 1331–1359.

5 Davidsson, P. and Boman, M. (2000). A multi-agent system for controlling intelligent buildings. *The Fourth International Conference on MultiAgent Systems (ICMAS-2000)* (10–12 July 2000), p. 377. Boston, MA: IEEE Computer Society.

6 Qiao, B., Liu, K., and Guy, C. (2006). A multi-agent system for building control. *The IEEE/WIC/ACM international conference on Intelligent Agent Technology* (IAT'06), Hong Kong (18–22 December 2006), pp. 653–659. Hong Kong: IEEE Computer Society.

7 de Farias, C.M., Soares, H., Pirmez, L. et al. (2014). A control and decision system for smart buildings using wireless sensor and actuator networks. *Transactions on Emerging Telecommunications Technologies* 25 (1): 120–135.

8 Guerrieri, A., Fortino, G., and Russo, W. (2014). An evaluation framework for buildings-oriented wireless sensor networks. *Proceedings of the 14th IEEE/ACM International Symposium on Cluster, Cloud and Grid Computing*, Chicago, pp. 670–679 (26–29 May 2014).

9 Akkaya, K. and Younis, M. (2005). A survey on routing protocols for wireless sensor networks. *Ad Hoc Networks* 3: 325–349, 5.

10 Guerrieri, A., Geretti, L., Fortino, G., and Abramo, A. (2013). A service-oriented gateway for remote monitoring of building sensor networks. *Proceedings of the 2013 IEEE 18th International Workshop on Computer Aided Modeling and Design of Communication Links and Networks (CAMAD 2013)*, pp. 139–143 (September 2013).

11 Guerrieri, A., Fortino, G., Ruzzelli, A., and O'Hare, G. (2011). A WSN-based building management framework to support energy-saving applications in buildings. *Advancements in Distributed Computing and Internet Technologies: Trends and Issues*. Ch. XII, pp. 1–14. Hershey, PA: IGI Global.

12 Best Practice: Integrating and monitoring heterogeneous technology systems. the NYC Global Partners' Innovation Exchange website. http://www.nyc.gov/html/unccp/gprb/downloads/pdf/Tel%20Aviv_NETA.pdf (accessed 12 June 2017).

13 Buddhikot, M., Chandranmenon, G., Han, S. et al. (2003). Integration of 802.11 and third-generation wireless data networks. *The Twenty-Second Annual Joint Conference of the IEEE Computer and Communications. IEEE Societies (INFOCOM 2003)*, San Francisco, USA (30 March–3 April 2003).

14 Stanic, M., Mitic, D., and Lebla, A. (2012). A mobile agents framework for integration of legacy telecommunications network management systems. *Przeglad Elektrotechniczny* 88 (6), pp. 337–341.

15 Baker, P.C., Catterson, V.M., and McArthur, S.D.J. (2009). Integrating an agent-based wireless sensor network within an existing multi-agent condition monitoring system. *15th International Conference on Intelligent System Applications to Power Systems (ISAP'09)*, Curitiba, Brazil (8–12 November 2009).

16 Mesjasz, M., Cimadoro, D., Galzarano, S. et al. (2012). Integrating Jade and MAPS for the development of agent-based WSN applications. *The 6th International Symposium on Intelligent Distributed Computing (IDC 2012)*, Calabria, Italy (24–26 September 2012).

17 Chiti, F., Fantacci, R., Archetti, F. et al. (2009). An integrated communications framework for context aware continuous monitoring with body sensor networks. *IEEE Journal on Selected Areas in Communications* 27 (4): 379–386.

18 Fortino, G., Gravina, R., and Guerrieri, A. (2012). Agent-oriented integration of body sensor networks and building sensor networks. *Proceedings of 2012 Federated Conference on Computer Science and Information Systems (FedCSIS 2012)*, Wroclaw, Poland (9–12 September 2012), pp. 1207–1214.

19 Bellifemine, F., Poggi, A., and Rimassa, G. (2001). Developing multi agent systems with a FIPA-compliant agent framework. *Software Practice and Experience* 31: 103–128.

20 Chipara, O., Lu, C., Bailey, H.C., and Roman, G.-C. (2010). Reliable clinical monitoring using wireless sensor networks: experience in a step-down hospital unit. *8th ACM Conference on Embedded Networked Sensor Systems (SenSys 2010)*, Zurich, Switzerland (3–5 November 2010).

21 Rashidi, P. and Mihailidis, A. (2013). A survey on ambient-assisted living tools for older adults. *IEEE Journal of Biomedical and Health Informatics* 17 (3): 579–590.

22 Wang, L., Gu, T., Chen, H. et al. (2010). Real-time activity recognition in wireless body sensor networks: from simple gestures to complex activities. *The 16th International Conference on Embedded and Real-Time Computing Systems and Applications, ser. RTCSA'10*, Macau, China (23–25 August 2010), pp. 43–52. IEEE Computer Society.

23 Bellifemine, F., Fortino, G., Giannantonio, R. et al. (2008). Development of body sensor network applications using SPINE. *The 2008 IEEE International Conference on Systems, Man, and Cybernetics (SMC 2008)*, Singapore, (12–15 October 2008).

24 Giannantonio, R., Gravina, R., Kuryloski, P. et al. (2009). Performance analysis of an activity monitoring system using the SPINE framework. *The 3rd International Conference on Pervasive Computing Technologies for Healthcare, ser. Pervasive Health 2009*, London, UK (1–3 April 2009), pp. 1–8. IEEE Press.

9

Integration of Wearable and Cloud Computing

9.1 Introduction

As widely discussed so far, wearable sensors and BSNs provide a platform for many human-centered applications, ranging from health care to gaming, sports performance analysis, and social networking. There is currently an enormous public interest in biomedical sensor-based systems and wearable consumer electronics that allow individuals, ranging from children to elders, to monitor their health and control their fitness. In all BSN scenarios, assisted livings are monitored by BSNs to gather data streams for processing them in real time [1] and archiving them in remote data repositories for off-line analysis. Such scenarios imply that a huge amount of data could be transmitted, stored, and analyzed. Thus, such huge amount of data generated by BSNs requires a powerful and scalable processing and storage platform that is able to support both online and off-line analysis of sensor data streams. This chapter therefore provides a research-oriented perspective on the integration of wearable and cloud computing to fulfill the aforementioned requirement. After providing some basic elements on cloud computing and introducing the motivations and the challenges of integrating wearable computing and cloud computing, the chapter focuses on the virtualization of body sensor networks (BSNs) through a reference cloud-based architecture. We will then discuss and compare the state-of-the-art about WSN and BSN virtualization with respect to the features of such reference architecture. Finally, the chapter presents BodyCloud, a cloud-assisted BSN architecture for the development of community BSN applications. A set of diverse large-scale community BSN applications that can be engineered through BodyCloud is also discussed.

Wearable Computing: From Modeling to Implementation of Wearable Systems Based on Body Sensor Networks, First Edition. Giancarlo Fortino, Raffaele Gravina, and Stefano Galzarano.
© 2018 John Wiley & Sons, Inc. Published 2018 by John Wiley & Son, Inc.

9.2 Background

9.2.1 Cloud Computing

Cloud computing can be defined as a computing paradigm that is based on sharing computing resources rather than having local servers or personal devices to handle applications. Cloud computing is similar to grid computing [2], a computing paradigm where unused processing cycles of all computers in a network are harnessed to solve problems too intensive for any stand-alone machine. Cloud computing [3] thus provides flexible, robust, and powerful storage and computing resources, which enables dynamic data integration and fusion from multiple data sources. Moreover, a cloud computing-based approach can offer flexibility and adaptability in the management and deployment of data analysis workflows. The dynamic deployment of software components as cloud computing-based services removes the need for new client applications to be developed and deployed when the user requirements change. This also motivates and introduces an intrinsic competitive environment for the development and deployment of better services.

Cloud computing layers (Infrastructure as a Service – IaaS, Platform as a Service – PaaS, and Software as a Service – SaaS) and software components (e.g. databases and data mining workflow tools) can be customized to support a distributed (quasi) real-time system for the monitoring and analysis of BSN data streams.

Figure 9.1 shows the diagram of the cloud computing ecosystem. The cloud computing Provider exports the IaaS integrated with a Data Mining development environment as a PaaS to the Application Workflow Developer. The Workflow Developer deploys a particular application as SaaS to the End User (e.g. the cardiovascular doctor collecting sensor data from many patients or the medical staff at the health-care point gathers vital parameters from assisted livings). The front-end of the application can be developed, for example, for a mobile device to ensure mobility and portability. The approach can be based on the customization of an open-source cloud computing toolkit (e.g. Google App Engine – GAE, MS Azure, and Amazon EC2) using cloud computing standards [4] and integrated with well-known data mining development tools and workflow management systems (e.g. KNIME [5], RapidMiner [6], and Weka [7]).

9.2.2 Architectures for Sensor Stream Management

Data stream management systems (DSMS) [8–10] are designed to provide quick response time when managing large volumes of (time-dependent) data streams, e.g. sensor observations. DSMS employ window-based data processing combined with synopsis to process large volumes of time-dependent data. Using synopsis helps a DSMS in reducing the response time to queries. Global

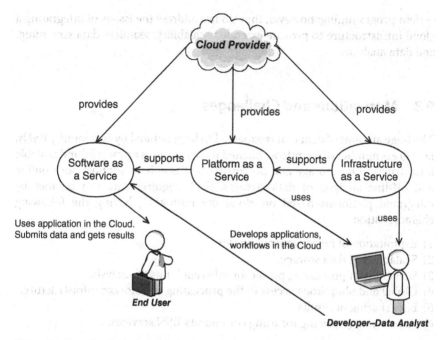

Figure 9.1 The cloud computing ecosystem.

Sensor Network (GSN) [11], TelegraphCQ [12], Aurora [13], and Stream [14] are some of the well-known proposals in the DSMS domain.

There exist several research projects to provide access, query, streaming, and management of WSN data. The Sensor Web project [15] provides a dynamic infrastructure that allows users to access sensor networks and the data streams generated from them. Sensor Information Networking Architecture (SINA) [16] is a middleware for querying, monitoring, and tasking of sensor networks. Tiny Application Sensor Kit (TASK) [17] is built on top of TinyDB, the well-known distributed database based on TinyOS [18], to provide high-level metadata management, query configuration, monitoring, and data visualization. These systems are appealing as they address the challenges related to large-scale (wireless) sensor resources and data sharing.

In recent years, there have been an increasing number of research studies to design and implement distributed platforms based on BSNs for e-Health applications. Many national and international research projects in academia, industry, and government focus on the development and deployment of health-care platforms in which wearable sensors are attached to patients for enabling 24/7 monitoring of vital parameters. Examples of such projects include CodeBlue [19], DexterNet [20], SPINE [1, 21, 22], SPINE2 [23–25], and Titan [26]. These systems provide effective programming abstractions atop the low-level TinyOS system programming; however, they do not address the issues of integrating a cloud infrastructure to provide extended scalability, seamless data streaming, and data analysis.

9.3 Motivations and Challenges

The huge amount of data that is expected to be generated by community BSNs, i.e. a great number of (semi)coordinated BSNs, requires a powerful and scalable infrastructure for storage and processing that is able to support both online and off-line analysis of data streams. Such requirements can be met by integrated platforms based on cloud computing [3] having the following characteristics:

1) Exploitation of heterogeneous sensors.
2) Scalability of data storage.
3) Scalability of processing power for different kinds of analysis.
4) Global and ubiquitous access to the processing and storage infrastructure.
5) Easy sharing of results.
6) *Pay-as-you-go* pricing for using community BSN services.

The integration of BSNs with cloud computing can provide important benefits in the following four main aspects:

- *Management*: BSN data management deals with the fundamental task of defining how BSN data streams are efficiently collected, managed, stored, and conveyed for final processing. Activities associated with the collection and management of data feeds from BSNs in real time may be distributed in time and/or space [27]. Time distribution refers to activities taking place at different times, while being coordinated to have a coordinated effect, such as in a workflow. Space distribution implies that activities may take place at different locations, while such activities are interconnected by a network. A cloud computing infrastructure can ease the management of distributed data and processes and support advanced functionalities such as information fusion at different levels (sensor, processed data, and decision) [28].
- *Processing*: the data streams collected from BSNs are processed and (sometimes) combined into measurement composites, e.g. combining body temperature readings and blood pressure into a health chart for given assisted livings. In the presence of numerous incoming data streams from a set of BSNs, in order to make critical decisions in real time, BSN data processing requires fast processing that may be computing and/or resource intensive. Harnessing the computational resources of a cloud computing infrastructure can be performed for the required provisioning of computing resources [3].

- *Service composition and invocation*: BSN-processed data are usually associated with meaning, confidence, and quality information. Specifically, the data are associated with information on how they were processed (derivation), for whom and why they were collected (agency), and how they may be distributed (rights). This process can be modeled and executed through automatic formation of workflows and invocation of services. It can be fully supported by a platform based on a cloud computing infrastructure.

- *Analysis*: BSN datasets are imported into analysis tools and modeling is further performed for the use in various applications and decision-making systems. The analysis activity depends on suitable storage and middleware technologies to perform highly swift data processing. It can be fully supported by using the processing power of cloud computing infrastructures that provide fast response times.

While there are main advantages of BSN adoption in various applications, there are a number of associated challenges that need to be addressed [29]. Moreover, the integration of BSNs with a cloud computing infrastructure raises additional challenges related to data management, system implementation, and real-time computing.

In the following, we first list BSN-related challenges and then we discuss specific challenges regarding BSN-Cloud computing systems that integrate BSN with cloud computing to perform effective data stream processing.

9.3.1 BSN Challenges

- *Interference reduction*: BSNs use wireless connectivity for communications. The BSN system should be able to reduce/mitigate interference on the wireless link and increase the co-existence of wearable sensor nodes with other networked devices [30]. This is important to ensure that the functionalities of BSN nodes (and the whole BSN system) do not degrade due to the presence of other devices capable of possible interruption/interference in the data transmission.

- *Data validation and consistency*: data collected from multiple sensor nodes need to be collected and analyzed seamlessly. BSN sensors are subject to inherent hardware, network, and communication failures that may result in erroneous gathered datasets [31]. It is crucial that the sensed data are validated and data quality is maintained under control to reduce any noise in the data and identify possible weaknesses in the BSN system.

- *Heterogeneity and interoperability*: a BSN system should be capable of integrating various different sensors in terms of complexity, power efficiency, storage, and ease-of-use [20]. Moreover, it should provide a common interface between the sensors and a storage service to facilitate remote storage and viewing of sensed data as well as access to external processing and networked analysis tools [32].

Moreover, a BSN system requires ensuring seamless data transfer across different standards to promote information exchange, plug-and-play device interaction and uninterrupted connectivity [33].

- *Security and privacy*: transmission of BSN data streams should be secured to prevent potential intruders [34]. Moreover, integrity of each assisted living's data has to be maintained with guarantee that one assisted living's data is not mixed with another assisted living's data. Another key problem of BSN users is to protect the privacy of personal data [35]. A BSN system should ensure that assisted livings' privacy is maintained even when data is being analyzed using an external tool.
- *Programming*: BSNs are usually programmed by using the low-level APIs provided by the adopted BSN sensor platforms (e.g. TinyOS and ZigBee). However, to enable a more rapid and effective prototyping, higher level programming abstractions offered by a BSN middleware are needed [1].

9.3.2 BSN/Cloud Computing Integration Challenges

- *Interfacing BSNs with cloud computing infrastructures*: a well-defined interface between BSN resources and the cloud fabric needs to be established. Communication interfaces are in fact required to manage network connectivity between BSN and the cloud. BSN nodes could be exposed as cloud services and indexed via indexing services according to functions/ services they are able to provide. Moreover, the presence of provision is important to manage sensing jobs and data streams from the sensor network. The key technology is therefore virtualization. Finally, an integration framework should provide various services for the underlying wearable sensor resources such as power management, security, availability, and QoS.
- *Data stream management*: data management includes data format conversion into standard formats (when available), data cleaning and aggregation to improve data quality, and data transfer to storage clouds.
- *Complex event processing*: real-time data streams from single or multiple BSNs may trigger certain events and services in the cloud. These data streams are analyzed through complex event processing (CEP) algorithms and the results are used in applications for decision making by identifying contextual and situational information.
- *Massive scale and real-time processing*: integration of even heterogeneous BSNs generating vast amounts of data is a challenge, especially in the presence of real-time requirements. BSNs generating real-time multimedia content, such as streaming video, audio, and images, pose additional issues in order to accurately process and store the data in a cloud environment.
- *Large-scale computing frameworks*: the allocation of computational and storage resources as well as data migration in the cloud is critical when BSN data

sources are not colocated. This is particularly challenging when the datasets and their corresponding access/search services are geographically distributed within the cloud.

- *Harvesting collective intelligence*: while heterogeneous and real-time BSN data feeds allow improving decision making by using data- and decision-level fusion techniques, maximizing the intelligence that can be exploited from massively colocated information in the cloud is challenging.
- *Large-scale application development*: the development of large-scale BSN systems is a complex task that needs suitable and effective software engineering methodologies and tools. Specifically, an application needs to be designed at a high level of modeling abstraction, implemented according to a given methodological approach, and then seamlessly deployed onto the target cloud platform using suitable tools.

9.4 Reference Architecture for Cloud-Assisted BSNs

A general reference architecture for the integration of BSNs and cloud computing is portrayed in Figure 9.2.

This architecture is supported by the following requirements:

- Efficient collection of sensor data streams from highly decentralized BSNs.
- Effective management of sensor data streams.
- Configuration of a scalable framework to support processing of multiple sensor data streams for (even concurrent) application services.
- Persistent storage and exchange of sensor data and analysis results to enable further decision-making.
- Workflow-oriented decision-making applications dynamically developed through distributed services/components' mash-up.
- Advanced visualization services (both for raw and processed sensor data, and for analysis results) that can be flexibly customized by the final users.
- Multiple-level security for at least sensor data collection (from sensors to the coordinator), sensor data transmission (from the coordinator to the cloud), and data analysis/visualization services (cloud access).

Each requirement is discussed in detail in Sections 9.4.1–9.4.7.

9.4.1 Sensor Data Collection

Sensor data collection allows for capturing sensor readings from the BSN sensor nodes, converting the raw values to meaningful measurements, or directly using the preprocessed data and store (annotated) data as necessary. A transport layer is used to assist in collecting sensor data points across a large dimension in (quasi) real time. Usually such data acquisition is deployment dependent.

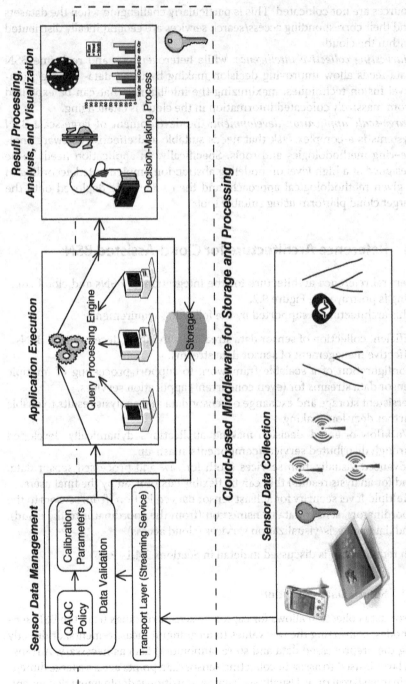

Figure 9.2 Reference architecture for the integration of BSN and cloud computing.

For TinyOS sensor platforms, TinyOS SerialForwarder (for TinyOS 1.x and 2.x compatible motes) can be used to capture raw data directly from remote sensors. There can also be hardware-specific proprietary APIs to read raw sensor readings directly from BSN sensors. Indeed, BSN middlewares are currently available for such purpose: CodeBlue [19], Titan [26], RehabSPOT [36], and particularly SPINE [1, 21, 22] and SPINE2 [23–25] provide high-level abstractions and mechanisms to capture, (pre)elaborate, and transmit sensor data to static and mobile base stations. In mobility scenarios, a mobile device (also called mobile coordinator) is interposed between the BSN and the cloud platform. The mobile coordinator collects sensor data from the BSN and transmits them onto the cloud platform. For instance, Android-SPINE, the Android version of the SPINE middleware [1], can be used to enable Android-based mobile devices, such as smartphones and tablets, to be the BSN mobile coordinator. In particular, data collected through Android-SPINE can be easily streamed up to the cloud side through an Internet-based connection. In Android-SPINE, wearable sensors' communication is currently based on Bluetooth.

9.4.2 Sensor Data Management

After data collection, data are passed through a data calibration process to ensure the validity and consistency of the gathered sensor data stream. A Quality Assurance Quality Control (QAQC) framework, comprising statistical models, can be applied to perform outlier detection, missing data handling, aggregation, detection of measurement changes, automated data correction, and, if needed, data compression in streaming sensor data [37]. In particular, data calibration APIs should be provided to support the implementation of custom calibration functions or third-party data calibration packages should be reused. Indeed, data quality can also be checked at the sensor data collection side, at the sensor node side [38] (see Chapter 5), and/or at the BSN coordinator side. When a calibrated data stream is available, it is exposed to the application services executing in the cloud and also stored (with metadata annotation providing meaning of the sensor streams) in the storage cloud resources for future use. Having such components to deal with a large number of sensor streams arriving continuously from numerous sensors, the cloud-enabled system should provide full support to guarantee reliability and robustness.

9.4.3 Scalable Processing Framework

Application services (e.g. ECG data analysis, health monitoring, sports performance monitoring, and rehabilitation control) are hosted in the VM-based cloud computing infrastructure for application execution. The communication between calibrated data streams and the cloud infrastructure (i.e. between *sensor data management* and *application execution* components) should be done through the use of nonblocking callback APIs. These APIs should allow application services to receive calibrated sensor data streams as streams arrive into the system. As applications (or services) are executed inside a VM, a data connection is required to transmit the results of the experiment to the *result-processing* component. The APIs should be able to buffer data streams within a time window in case the application service does not respond or the call back connection is lost. Thus, using persistent buffers in the cloud system to communicate between the BSNs and the hosted application services would ensure users from any potential data loss. The output produced by the applications is transmitted to generate continuous data streams incorporating the results and also stored in the persistent storage.

9.4.4 Persistent Storage

The cloud-enabled storage component is fundamental for a cloud-assisted BSN architecture in order to persistently store data coming from (i) the sensor data collection process, (ii) the processed data streams, and (iii) the data analysis results. Such time-dependent datasets can be therefore reused either online or off-line.

The persistent storage component is characterized by the following elements:

- *Storage virtualization*, which refers to thin provisioning of the storage cloud infrastructure, with the assistance of a management software layer, to automate data availability and security management. Storage virtualization, which can be encapsulated in an orchestrated workflow, assists in persistency and optimization of existing storage, and in provision of new storage.
- *Enterprise resource management* in order to reduce administrators' efforts to manage heterogeneous storage cloud infrastructures. Based on the administrator's policies, the management software in the cloud-based BSN gathers information for managing the storage environment.
- *Hierarchical storage management* through a tiered storage infrastructure to manage growth and provide different levels of service to BSN users. It is used for storage space management through automatic data migration and transparent data restore in failure situations.
- *Archive management* to provide BSN data retention over time as the stored data grows. Storage archives copy data for a dedicated time frame, defined by the cloud-based BSN administrator's policies.
- *Recovery management* is the ability to recover backup/archived data, thus ensuring effective operational continuance of sustained performance. Recovery management assists in recoverability in a heterogeneous cloud storage environment.

- *Interfacing APIs* to interact with different components of the cloud-based BSN architecture. The exposed APIs allow the abstraction of complex functionalities, feed input to application execution, data transfer in and out of storage, and runtime interactions.

Moreover, cloud-based BSN architectures can use Google Bigtable [39] or Azure BLOB [40] storage. These cloud storage services allow managing large-scale structured data across thousands of commodity servers, ensuring persistent data management and fulfilling latency requirements.

9.4.5 Decision-Making Process

Upon the availability of outputs from the processing stage, the result-processing service/component informs internal (user-programmed) or external decision-making processes (by reusing existing tools) about specific situations. This component can provide a set of user-defined policies that are specific to particular BSN scenarios. Furthermore, a client *decision-making process* application can register with the result-processing component to submit continuous query for gathering continuous delivery of latest results. With the use of a continuous query, a client application can specify the window size (i.e. the amount of data used at the processing stage) and the sliding predicate (i.e. how frequent a continuous query is to be evaluated). The decision-making process is usually workflow-oriented: it is performed through automatic formation of workflows and invocation of services. Such operational workflow requires a platform to support automatic workflow formation and service invocation, potentially through a cloud infrastructure.

9.4.6 Open Standards and Advanced Visualization

Open standards for data and for workflow definitions allow input and intermediary data to be propagated through processing elements in data analytics and mining workflows. They also allow the workflow components to be exchanged and executed in distributed environments. For example, the Attribute-Relation File Format (ARFF) [41] is an ASCII text file format that describes a list of instances sharing a set of attributes. The data-flow programming paradigm adopted in KNIME workflows [5] is based on an XML-based workflow specification format and on an intermediary data format that incorporates rich metadata information about the data attributes. The Predictive Model Markup Language (PMML) [42] is an XML-based open standard for the description and exchange of models produced by data-mining algorithms and for data manipulation and transformations.

However, there is no open standard for the representation and visualization of the data analysis results. A powerful visualization service is necessary, as the cloud computing environment stores and processes enormous amounts of data. The visualization service should provide various predefined and user-defined views on the data and analysis results. The visualizations and views can be implemented with heterogeneous languages like XML, OLAP/data warehouse tools, and/or specific graphical languages/frameworks. Separating the formal specifications of the visualization from the graphical primitives used to generate the views in a given client application is an important aspect for a cloud-based distributed environment with a wide heterogeneity of supported devices.

9.4.7 Security

Considering social, ethical, and legal aspects of human-centered systems such as BSN systems, data in cloud-based BSNs (i.e. data collected from BSNs, and stored and processed/analyzed in the cloud) are highly sensitive and should be managed properly to guarantee people privacy [35].

It is therefore crucial to define system-wide security mechanisms to guarantee confidentiality, data integrity, as well as fine-grained access control to data and services.

We devise a three-level security framework for cloud-based BSNs:

- *Sensor data collection level*: securing data communications from sensors to the BSN coordinator through encryption. Wearable sensor nodes have limited computing and energy resources, and encryption consumes time and energy, so specialized in-node hardware needs to be exploited (e.g. 128-bit AES encryption hardware is included in the TelosB sensor platform).
- *Sensor data transmission level*: from the BSN coordinator to the cloud. Data streams can be transferred onto the cloud through Transport Layer Security (TLS)/Secure Sockets Layer (SSL), which is a proven technology. However, new security mechanisms dealing with mobility need to be purposely defined.
- *Sensor data management and access level*: managing and accessing data and services on the cloud. Data stored and processed in the cloud computing infrastructure need to be protected by authentication and authorization measures, and can also be encrypted, if needed. Moreover, the cloud services used by different actors of the system need to be secured through specific access control policies.

Finally, as cloud-based BSNs can support different application domains (from health care to crowdsourcing), specific national or transnational security/privacy standards, e.g. normative on medical data treatment, processing, and storing, should be introduced at the application level.

9.5 State-of-the-Art: Description and Comparison

The integration of WSNs/BSNs with large-scale distributed computing infrastructures is a recent research area attracting both academia and industry researchers. A few interesting works have been to date proposed. In the following, we first describe solutions integrating WSNs and cloud computing; then, we discuss specific infrastructures that integrate BSNs and cloud computing towards cloud-based BSNs.

9.5.1 Integration of WSNs and Cloud Computing

In Ref. [43], a SaaS architecture for sensor network analytical services is proposed. It is implemented atop a PaaS layer (e.g. GAE and MS Azure) and is organized in three layers: (i) sensor data management, which collect sensor data streams coming from the WSN gateway; (ii) run-time for filter analysis, which supports the execution of processing workflows for sensor data according to the pipe-and-filter paradigm; and (iii) filter management, visualization, and notification, which are three components that respectively allow for the definition and management of the processing filter chain, for the visualization of analyzed data, and for the notification of events to external applications.

The authors in Ref. [44] propose the Open Sensor Web Architecture (OSWA). OSWA is an OGC (Open Geospatial Consortium) Sensor Web Enablement standard-compliant software infrastructure for providing service-oriented-based access to and management/integration of sensors. OSWA also integrates emerging distributed computing platforms such as SOA and Grid Computing. OSWA is designed around the conventional Grid layers: Fabric, Services, Development, and Application. Specifically, the OSWA-based platform provides a number of sensor services such as sensor notification, collection, and observation; data collection, aggregation, and archive; sensor coordination and data processing; faulty sensor data correction and management; and sensor configuration and directory service.

In Ref. [45], the authors propose a new infrastructure, called Sensor-Cloud, which can manage physical sensors on an IT infrastructure for sensors' virtualization. The Sensor-Cloud Infrastructure virtualizes a physical sensor as a virtual sensor on the cloud computing platform. Dynamic grouped virtual sensors on cloud computing can be automatically provisioned when the users need them through a portal server interacting with the workflow-oriented provisioning server, performing resource management, and a monitoring server, monitoring real/virtual sensors.

SAaaS [46] is a cloud-enabled SaaS architecture aiming at the management of wireless sensor and actuator networks (WSANs). SAaaS is a software stack that implements the following main functionalities: involvement of (W)SNs, smartphones, or other devices endowed with sensors and/or actuators, and their enablement for interoperation and management in a cloud environment; exploitation of volunteer-based methods for node involvement; functions and interfaces for federating SAaaS Clouds, either volunteer-based or commercial/ institutional.

The aforementioned works mainly describe architectural models and/or case studies and somehow identify related development issues. However, there is still a gap to fill in order to develop a cloud-based infrastructure that is targeted to BSN applications as the one proposed in Section 9.4. The research works discussed in Section 9.5.2 aim at the fulfillment of such a gap.

9.5.2 Integration of BSNs and Cloud Computing

In Ref. [47], the authors propose the development of an autonomic cloud environment for hosting an ECG data analysis service. In particular, they propose an autonomic cloud environment that collects people's health data and stores them to a cloud-based information repository and facilitates analysis on the data using software services hosted in the cloud. To evaluate the software design, a prototype system is developed, which is used as an experimental testbed on a specific use case, namely, the collection of electrocardiogram (ECG) data obtained at real time from volunteers to perform basic ECG beat analysis. The ECG software is hosted as a web-service such that any client-side implementation can simply call the underlying functions (analyze, upload data, etc.) without having to go through the complexities of the underlying application. The PaaS layer controls the execution of the software using three major components: (i) Container scaling manager, (ii) Workflow Engine, and (iii) Aneka Cloud middleware.

In Ref. [48], a secure and scalable cloud-based architecture for e-Health WSNs is proposed. The aim is to support (i) body sensor data collection from patients both hospitalized and at home and (ii) medical data management for e-Health monitoring. Collection is based on BSNs worn by patients and mobile/static devices working as Internet-based gateways. A cloud infrastructure is used for storing and retrieving the collected BSN data. Security protocols and mechanisms are defined to provide data security.

In Ref. [49], a cloud-assisted WBAN is proposed, specifically designed for pervasive health care in home, hospital, or outdoor environment. This system is composed of four main components: WBANs, wired/wireless transmission, cloud services, and users. WBANs can be based on fixed networks at home and on mobile devices (smartphone/tablet) at hospital and outdoor. Data are sent onto (public and private) cloud, providing several services (automatic diagnosis and alarm, location-based services, GIS services, real-time monitoring of patients, and medical knowledge sharing). Users can access the cloud according to their role and they are connected through social networks.

Finally, BodyCloud [50, 51] is a novel cloud-enabled system architecture that integrates BSNs' services with a cloud computing infrastructure. In particular, BodyCloud is a SaaS architecture that supports the storage and management of sensor data streams generated by SPINE-enabled mobile BSNs and the processing and analysis of the stored data using software services hosted in the cloud. BodyCloud endeavors to support several cross-disciplinary applications and specialized processing tasks. It enables large-scale data sharing and collaborations among users and applications in the cloud and delivers cloud services via sensor-rich mobile devices. BodyCloud also offers workflow-oriented decision support services to take further actions based on the analyzed BSN data. BodyCloud is fully compliant with the reference architecture described in Section 9.4.

9.5.3 A Comparison

In Tables 9.1 and 9.2, the main available architectures integrating WSNs or BSNs with a cloud computing platform are compared with respect to the requirements identified in Section 9.4:

- *Sensor Data Collection*: although, sensor data collection is provided by all architectures and is based on a (static and/or mobile) gateway device that gathers data from the body-worn sensors and transmits them to the cloud through an Internet-based connection, the exploited technologies are different at application, protocol, and system level. It is worth noting that SAaaS uses a complex software framework at the gateway side called Hypervisor, which is able to manage not only sensor reading collection but also to control actuator devices.
- *Sensor Data Management*: it is based on different paradigms (data-driven pipes and filters, rule-based planning, virtual sensors, and workflow-oriented). However, SAaaS, ECGaaS, Cloud BAN e-Health, and Cloud-Assisted WBAN do not specify any sensor data management paradigm.
- *Processing Framework*: it is basically the execution engine of the sensor data management paradigm carried out at the SaaS level or at the PaaS level. CC-WSN, SAaaS, Sensor-Cloud, and BodyCloud provide a processing framework at the SaaS level supported by a specific PaaS. The processing framework of OSWA and ECGaaS are implemented at the PaaS level. Finally, Cloud BAN e-Health and Cloud-Assisted WBAN do not support any specific processing framework.
- *Persistent Storage*: all architectures provide cloud storage but OSWA and Sensor-Cloud, which are based on stand-alone databases, and SAaaS, which does not specify the use of persistent storage.
- *Decision-Making Process*: it is fully supported only by BodyCloud through a flexible and distributed workflow-oriented model.

Table 9.1 Architectures for the integration of wireless sensor networks with cloud computing: a comparison.

	CC-WSN [43]	OSWA [44]	Sensor-cloud [45]	SAaaS [46]
Sensor data collection	WSN gateway based on HTTP/AJAX	Static gateway based on WSDL/SOAP	WSN gateway based on TCP/IP	Gateway node based on the Hypervisor framework
Sensor data management	Pipes, filters, and filter chain paradigm	Rule-based planning	Virtual sensors	N/A
Processing framework	Pipe, filters, and filter chain run-time engine at the SaaS level (GAE or MsA is the PaaS level)	Scheduler for plan execution	Workflow engine for service provisioning	Runtime split between the Hypervisor and the Cloud side
Persistent storage	Bigtables (provided by GAE) or BLOBs (provided by MsA)	Stand-alone database	Stand-alone database	N/A
Decision-making process	Not supported, delegated to external tools	N/A	N/A	N/A
Visualization service	User-defined views on sensor data and analysis results	N/A	Raw data visualization	N/A
Security	N/A	N/A	N/A	N/A

Table 9.2 Architectures for the integration of body area networks with cloud computing: a comparison.

	ECGaaS [47]	Cloud BAN e-Health [48]	Cloud-assisted WBAN [49]	BodyCloud [51]
Sensor data collection	Internet-based mobile BSN coordinator	Internet-based static/ mobile gateway	Internet-based static/ mobile gateway	Mobile Android BSN coordinator based on HTTP/REST
Sensor data management	N/A	N/A	N/A	Workflow-oriented paradigm
Processing framework	Workflow engine based on the Aneka PaaS	N/A	N/A	Workflow engine at the SaaS level (GAE is the PaaS level)
Persistent storage	Cloud storage	Cloud storage	Cloud storage	Bigtables (provided by GAE)
Decision-making process	N/A	N/A	N/A	Workflow-oriented process
Visualization service	Specific to the provided case study	N/A	N/A	XML-based Views on sensor data and analysis results
Security	N/A	RSK/ABE-based encryption of data SSL-secured communications	Key management Encrypted storage	OAuth-based authentication to access the cloud services

- *Visualization Service*: a customizable *visualization service* is only provided by CC-WSN and BodyCloud. Both architectures allow the implementation of user-defined views on sensor data and analysis results. In particular, the BodyCloud architecture [51] proposes an approach that integrates XML-based specifications for input data and for output data and their visualization.
- *Security*: only Cloud BAN e-Health, Cloud-Assisted WBAN, and BodyCloud provide security mechanisms. BodyCloud is currently based only on the OAuth protocol supported by the GAE to access the cloud services. Cloud BAN e-Health delivers an effective security framework centered on (i) data encryption based on a hybrid RSK (Randomly generated Symmetric Key) and ABE (Attribute-Based Encryption) method supported by a Health Authority, which also enables fine-grained access control to data and (ii) on SSL-secured communications. Finally, Cloud-Assisted WBAN is based on key management and encrypted storage.

9.6 BodyCloud: A Cloud-based Platform for Community BSN Applications

The BodyCloud platform aims at integrating BSNs and cloud computing PaaS infrastructures.

In particular, the BodyCloud architecture, shown in Figure 9.3, consists of four main subsystems (or sides):

- *Body-side*: it is the subsystem that monitors the assisted living by means of a BSN and sends the collected data to the cloud through a Java-enabled computer (desktop, laptop, or nanocomputer like Raspberry Pi) and/or an Android-enabled mobile device. In particular, data acquisition is currently based on SPINE [1] for computers and Android-SPINE, the Android version of the SPINE middleware [1] (see Chapter 3), for mobile devices. In particular, Android-SPINE allows Android-enabled smartphones and tablets to be used as coordinators of the BSN. Data collected through SPINE or Android-SPINE are then streamed up to the cloud-side by using the real-time data feed modality (see Cloud-side in next point). In Android-SPINE, communication of wearable sensors with the BSN coordinator is based on Bluetooth, whereas in SPINE communication can be based either on IEEE 802.15.4 or Bluetooth. The following functionalities are provided by the application-level SPINE protocol [21]: sensor discovery, sensor configuration, in-node processing, BSN activation/deactivation, data collection, and logging. Finally, the current SPINE implementation fully supports IEEE 802.15.4 TinyOS sensor nodes and the Bluetooth-based Shimmer sensor nodes.
- *Cloud-side*: it is the subsystem that fully supports specific applications through data collection, processing/analysis, and visualization. In particular,

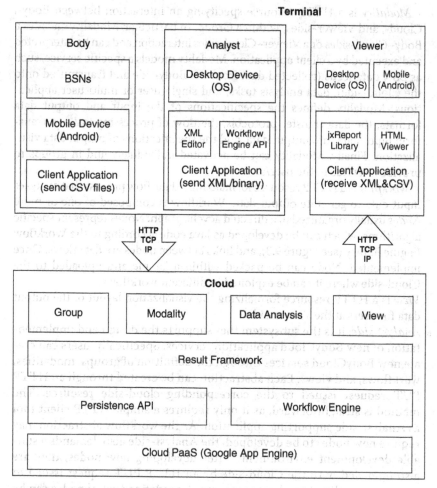

Figure 9.3 The BodyCloud architecture.

applications can be defined through four programming abstractions: *Group, Modality, Workflow,* and *View.*

Group is an HTTP resource formalizing an application manipulating a specific BSN data source. In particular, Group consists of three correlated subresources: (i) *Collector,* which is intended to collect BSN data that comply with the same data specification; (ii) *Data,* which represent the actual data collected by the Group and based on different formats (e.g. CSV, ARFF, and JSON); (iii) *Contributor,* which is a subresource containing the users who uploaded data to the Group. In particular, Data is grouped on a per user basis.

Modality is a HTTP resource specifying an interaction between Body-, Cloud-, and Viewer-sides, within a Group. In particular, Modality encodes a Body–Cloud-sides or a Viewer–Cloud-sides interaction and can be interpreted and executed by a client application. Modality models a specific service, such as BSN data feeds (collected data from the Body-side and transmitted onto the Cloud-side), data analysis tasks, and single-user or multi-user applications. Modality defines the specifications of the input and output data formats, the data transfer protocols, the flow of processing tasks to transform input data into output data, and the specifications of output data visualization. Finally, Modality can be activated individually and in groups to provide a service to the user/s.

Workflow is a HTTP resource formalizing a data-flow process that analyzes input data to generate output data. Workflow is composed of one or more *Nodes* usually organized in a directed acyclic graph. Nodes represent specific algorithms, which can be developed as Java code according to the Workflow Engine library (see Figure 9.3), and links between nodes are data flows. Once implemented, Node can be packed within a jar file and uploaded to the Cloud-side where it can be exploited in different workflows.

View is a HTTP resource formalizing the visualization layout of the output data for users at the Viewer-side.

- *Analyst-side*: it is the subsystem that supports the design and implementation of new BodyCloud application services. Specifically, users can create new BodyCloud services through the definition of groups, modalities, workflows, and views. Each abstraction can be created through an HTTP PUT request issued to the corresponding cloud-side resource. The method is straightforward, as it only requires a simple HTTP client tool as Analyst-side supporting application. As the workflow abstraction may require new nodes to be developed, the Analyst-side also demands a suitable development environment. After developing new nodes, they are also uploaded onto the Cloud-side by an HTTP PUT request issued to the corresponding Cloud-side resource. A predefined set of nodes can be easily made available, depending on the adopted implementation of the Workflow Engine.
- *Viewer-side*: it is the subsystem that visualizes the output produced by the data analysis through advanced graphical reporting facilities. The graphical view is automatically generated by applying the *View* specification (defined in the modality) to the output data. Specifically, as part of the current BodyCloud prototype, a Java library, named *jxReport*, was developed and integrated into the client application. The jxReport library provides functionalities to generate HTML reports from an XML schema and a data model, thus allowing the desirable separation between the data model and the view. During the graphical report generation, jxReport reads the model, e.g. from a CSV file, and draws the graphical elements specified in the XML

document based on the model data. The jxReport library is highly portable and can be used in any Java-based environment (e.g. mobile or desktop).

From an implementation viewpoint, Group, Modality, Workflow/Node, and View are supported by a RESTful web service (Server Servlet), implemented using the Restlet Framework, making the interaction with the Cloud-side fully based on the HTTP methods *get, put, post,* and *delete.* The interactions are authenticated by the OAuth Verifier component based on OAuth 2.0. The Cloud-side is supported by the GAE PaaS[1] that provides the Datastore API, atop which the Persistence Layer managing the collected BSN data is built, and the Task Queue API, which enables asynchronous execution of tasks triggered by requests.

9.7 Engineering BodyCloud Applications

BodyCloud supports an effective approach for the rapid prototyping of large-scale applications based on BSNs. A BSN service definition based on the BodyCloud approach can be developed and deployed on the basis of the following five phases organized as a workflow-based process in Figure 9.4:

1) *Development and upload of the processing/analysis algorithms*: design, implementation, and upload of any custom processing/analysis algorithms in terms of (processing/analysis) *nodes*. All uploaded nodes are stored into the Cloud-side and can be exploited by any BodyCloud user. Of course, this phase is optional as users can directly use algorithms already existing in the Cloud-side.
2) *Definition of the Data Source (or Group)*: definition of a Group containing the specification of the data that can be gathered from the BSN and then possibly processed by algorithms defined in phase 1 or already available in the Cloud-side.
3) *Definition of the Analysis Workflow*: definition of the data analysis process through the combination of the (uploaded and/or already uploaded) nodes

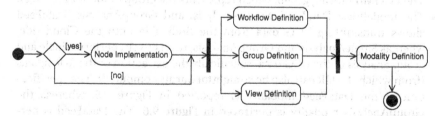

Figure 9.4 Workflow schema of the BodyCloud approach for developing community BSN applications.

---

1 https://cloud.google.com/appengine/

and their static parameters into a workflow. The starting node of the workflow should read the input data from the Data Source.

4) *Definition of the View*: definition of one or more graphical formats (or *views*) for the data produced by the processing/analysis workflow.

5) *Definition of the Modalities*: definition of at least a Body-side specific modality and a Viewer-side specific one. The Body-side modality should have an input data specification similar to the Group definition, an action that will upload the data to the group defined in phase 2 and no output specification. The Viewer-side modality should perform the workflow execution as action, the parameters of which must be defined accordingly to the node definition. Its output specification must match with the workflow output and contains the correlated reference to the view.

In the following subsections, we provide four BSN community applications supported by BodyCloud (ECGaaS, FEARaaS, REHABaaS, and ACTIVITYaaS).

9.7.1 ECGaaS: Cardiac Monitoring

The ECG as a Service (ECGaaS), which was developed by exploiting the BodyCloud approach, allows monitoring (collect, process, store, analyze, and visualize) ECG data coming from individuals or a group of people (e.g. assisted livings, athletes, and emergency teams). The ECG is the standard method for measuring the electrical and functional activity of the heart and is commonly used to diagnose cardiovascular diseases and cardiac abnormalities. In particular, in the developed application service, the ECG signal is captured by the Body-side, through a Shimmer sensor node equipped with the ECG board, and sent to the Cloud-side in which the R-R intervals and heart rate (HR) [52] are extracted through QRS-complex detector algorithms [53] deployed as nodes in the BodyCloud system.

The specific entities (group, modality, workflow, and view) defining the ECGaaS are:

- The ECGMonitoring group, which represents the group of monitored users.
- The modalities: DataFeed, SingleAnalysis, and GroupAnalysis. DataFeed allows transmitting ECG data from the Body-side onto the Cloud-side, whereas SingleAnalysis and GroupAnalysis, respectively, perform single and group analysis of the ECG data, specifically the extraction of the R-R signals (from which the HR can also be straightforwardly computed). The specification of the DataFeed modality is reported in Figure 9.5, whereas the GroupAnalysis modality is portrayed in Figure 9.6. The DataFeed is performed every 60s. The GroupAnalysis gets all the contributors (i.e. the identifiers of the involving participants) and executes the workflow on their data, thus providing the tachogram of all participants.
- The EcgToRR workflow (see Figure 9.7), which models a workflow composed of two sequential nodes able to read the collected ECG user data through the

```
<modality>
  <inputSpecification>
    <data>
      <name>ECGShimmerSample </name>
      <type>INTEGER</type>
      <source>ECGShimmerSensor</source>
    </data>
  </inputSpecification>
  <init-action>
    <uri>/group/ecg-monitoring/data</uri>
    <method>DELETE</method>
  </init-action>
  <action>
    <uri>/group/ecg-monitoring/data</uri>
    <method>PUT</method>
    <repeat>true</repeat>
    <trigger after="60"/>
  </action>
</modality>
```

Figure 9.5 ECGMonitoring DataFeed modality.

```
<modality>
  <init-action>
    <uri>/group/ecg-monitoring/contributors</uri>
    <method>GET</method>
  </init-action>
  <action>
    <uri>/engine/workflow/ecg</uri>
    <method>POST</method>
    <parameter>
      <name>sourceUser</name>
      <reference xpath="//users/user"/ type="MAP">
    </parameter>
    <parameter>
      <name>sourceGroup</name>
      <value>ecg-monitoring</value>
    </parameter>
    <repeat>false</repeat>
  </action>
  <outputSpecification>
    <data>
      <name>rr</name>
      <type>DOUBLE</type>
    </data>
    <view>/view/tachogram.xml</view>
  </outputSpecification>
</modality>
```

Figure 9.6 ECGMonitoring GroupAnalysis modality.

```
<workflow>
  <node>
    <type>UserDataReader</type>
  </node>
  <node>
    <type>RR</type>
  </node>
</workflow>
```

Figure 9.7 EcgToRR workflow.

(a)

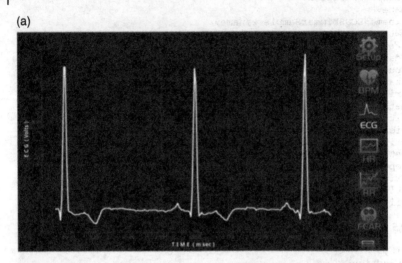

(b)

Figure 9.8 GUI view. (a) ECG wave plotting and (b) beat per minute instantaneous value.

data reader node and extract the R-R signal from the ECG data through the RR node.

- The Tachogram View, which is the graphical format through which the R-R signal will be rendered at the Viewer-side. The ECGaaS GUI, which is portrayed in Figure 9.8, allows visualizing the ECG plot and the HR ([bpm]) in real time.

9.7.2 FEARaaS: Basic Fear Detection

Apart from its common use in health care for the diagnosis of the cardiac status, the ECG signal can be exploited to detect emotions. The ECG is in fact very reactive to physiological responses due to emotions and other external factors. Other methods use facial recognition to detect/recognize emotions; however, they are invasive, as they require the placement of electrodes and cameras to detect subtle changes in the person's face. The advantage of using the ECG signal for detecting basic emotions is that a person can be monitored using noninvasive wearable cardiac sensors such as smart watches, sport electronic chest bands, or even smart textiles. A basic fear status (which is not yet cognitive fear, i.e. the response when a person is in danger) could be detected by analyzing the ECG signal. The basic cardiac physiological response that could generate the state of fear is the Cardiac Defense Response (CDR) [54]. On the basis of the algorithm for the CDR detection proposed in Ref. [53], a basic fear detection service (FEARaaS) was easily developed on BodyCloud, by also reusing some system components and entities defined for the ECGaaS.

The specific entities (group, modality, workflow, and view) defining the FEARaaS are:

- The CDRDetection group, which represents the group of monitored users.
- The ECGDataFeed (see Figure 9.9), SingleCDRAnalysis (see Figure 9.10), and GroupFearDetectionAnalysis modalities. ECGDataFeed is the same modality as in ECGaaS (see Section 9.7.1). SingleCDRAnalysis performs the CDR detection on a single subject and provides true if the CDR is detected, false otherwise. GroupFearDetectionAnalysis performs the CDR detection on a group and provides a positive result if the number of people having a CDR in a given time period exceeds a given threshold.
- The SingleCDR workflow (see Figure 9.11), which models a workflow based on three sequential nodes, is able to (i) read the collected ECG user data

```
<modality>
  <inputSpecification>
    <column>
      <name>heartbeat</name>
      <type>DOUBLE</type>
      <source>HEARTBEAT</source>
    </column>
  </inputSpecification>
  <init-action>
    <uri>/group/cdr</uri>
    <method>DELETE</method>
  </init-action>
  <action>
    <uri>/group/cdr</uri>
    <method>PUT</method>
    <repeat>true</repeat>
    <trigger after="10" />
  </action>
</modality>
```

Figure 9.9 CDRDetection DataFeed modality.

```
<modality>
  <init-action>
    <uri>/group/fear-detection/contributors</uri>
    <method>GET</method>
  </init-action>
  <action>
    <uri>/engine/workflow/cdr</uri>
    <method>POST</method>
    <parameter>
      <name>sourceUser</name>
      <reference xpath="//users/user"/>
    </parameter>
    <parameter>
      <name>sourceGroup</name>
      <value>cdr-monitoring</value>
    </parameter>
    <repeat>false</repeat>
  </action>
  <outputSpecification>
    <data>
      <name>cdr</name>
      <type>BOOLEAN</type>
    </data>
    <view>/view/cdrplot.xml</view>
  </outputSpecification>
</modality>
```

Figure 9.10 SingleCDRAnalysis modality.

```
<workflow>
  <node>
    <type>UserDataReader</type>
  </node>
  <node>
    <type>RR</type>
  </node>
  <node>
    <type>CDR</type>
  </node>
</workflow>
```

Figure 9.11 SingleCDR workflow.

through the data reader node, (ii) extract the R-R signal from the ECG data through the RR node, and (iii) apply the CDR detection algorithm to the R-R signal. An interesting enhancement is the GroupCDR workflow, which could be based on the SingleCDR workflow to which the node, which processes the group fear detection algorithm, has to be added.

- The CDR View allows to display the results provided by the (single or group) CDR detection. In Figure 9.12 the GUI at the Viewer-side, which displays the positive CDR detection, is portrayed.

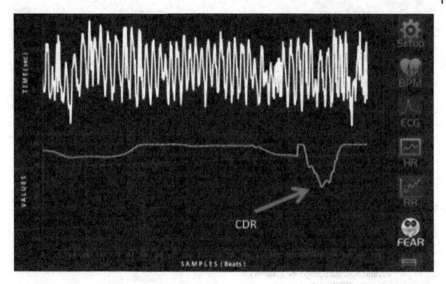

Figure 9.12 GUI view: detection of a CDR.

9.7.3 REHABaaS: Remote Rehabilitation

The remote rehabilitation application service (REHABaaS) involves remote rehabilitation of the limbs of assisted livings. Currently, the involved joints are elbows and knees. The service is based, on the Body-side, on two wearable sensor nodes equipped with 3-axial accelerometers. Sensors are placed in specific positions of the limbs for collecting accelerometer data, which are then processed by the BSN coordinator to provide specific rehabilitation information such as extension angles of elbows and knees [55].

The specific entities (group, modality, workflow, and view) defining the REHABaaS are:

- The Rehab Group represents the group of monitored users to be rehabilitated.
- The RehabDataFeed Modality (see Figure 9.13) allows transmitting the rehabilitation data from the Body-side to the Cloud-side.
- The Single RehabDataAnalysis Modality (see Figure 9.14) performs analysis of the single subject based on the RehabDataAnalysis workflow (see Figure 9.15) and provides statistics about the progress of the rehabilitation.
- The RehabData View, which is the graphical format through which the rehab data will be rendered at the Viewer-side. Figure 9.16 shows the web-based GUI for the knee rehabilitation: the exercise of the patient is compared with a reference exercise in terms of knee extension and inclination angles and thigh torsion.

```
<modality>
  <inputSpecification>
    <data>
      <sensor1Data>
        <name>AccXSample</name>
        <type>INTEGER</type>
        <source>ECGShimmerSensor1</source>
        <name>AccYSample</name>
        <type>INTEGER</type>
        <source>ECGShimmerSensor1</source>
        <name>AccZSample</name>
        <type>INTEGER</type>
        <source>ECGShimmerSensor1</source>
      </sensor1Data>
      <sensor2Data>
        <name>AccXSample</name>
        <type>INTEGER</type>
        <source>ECGShimmerSensor2</source>
        <name>AccYSample</name>
        <type>INTEGER</type>
        <source>ECGShimmerSensor2</source>
        <name>AccZSample</name>
        <type>INTEGER</type>
        <source>ECGShimmerSensor2</source>
      </sensor2Data>
      <extensionAngle>
        <name>AngleSample</name>
        <type>INTEGER</type>
        <source>BSN</source>
      </extensionAngle >
    </data>
  </inputSpecification>
  <init-action>
    <uri>/group/rehab-monitoring/data</uri>
    <method>DELETE</method>
  </init-action>
  <action>
    <uri>/group/rehab-monitoring/data</uri>
    <method>PUT</method>
    <repeat>true</repeat>
    <trigger after="1"/>
  </action>
</modality>
```

Figure 9.13 RehabMonitoring DataFeed modality.

9.7.4 ACTIVITYaaS: Community Activity Monitoring

ACTIVITYaaS is a BodyCloud service supporting real-time, noninvasive human activity recognition and monitoring. At the Body-side, it uses two wearable motion sensors and a personal mobile device where a graphical application provides instantaneous feedback to the user; in addition, when Internet connectivity is available, data are also sent onto the Cloud-side for long-term, multiuser data storage and processing. Finally, the Viewer-side allows for remote access to such information at authenticated and authorized users [56, 57].

```
<modality>
 <inputSpecification>
  <column>
   <name>foreNode-accX</name>
   <type>INTEGER</type>
   <source>GENERIC</source>
  </column>
  <column>
   <name>foreNode-accY</name>
   <type>INTEGER</type>
   <source>GENERIC</source>
  </column>
  <column>
   <name>backNode-accY</name>
   <type>INTEGER</type>
   <source>GENERIC</source>
  </column>
  <column>
   <name>backNode-accZ</name>
   <type>INTEGER</type>
   <source>GENERIC</source>
  </column>
 </inputSpecification>
 <action>
  <uri>/group/rehab-aaservice/data</uri>
  <method>PUT</method>
  <repeat>true</repeat>
 </action>
</modality>
```

Figure 9.14 Single RehabMonitoringAnalysis modality.

```
<workflow>
  <node>
    <type>UserDataReader</type>
  </node>
  <node>
    <type>Stats</type>
  </node>
</workflow>
```

Figure 9.15 RehabMonitoring workflow.

The specific entities (group, modality, workflow, and view) defining ACTIVITYaaS are:

- The ActivityMonitoring group represents the group of monitored users.
- The RawAccelerationDataFeed (see Figure 9.17), FeatureDataFeed, and ActivityDataFeed modalities, respectively, implement the following three operating modes:
  - *Full-Cloud*: the Body-side will only collect the raw data and send this straight to the Cloud-side. The Cloud-side will then do all required processing (i.e. feature extraction and classification).

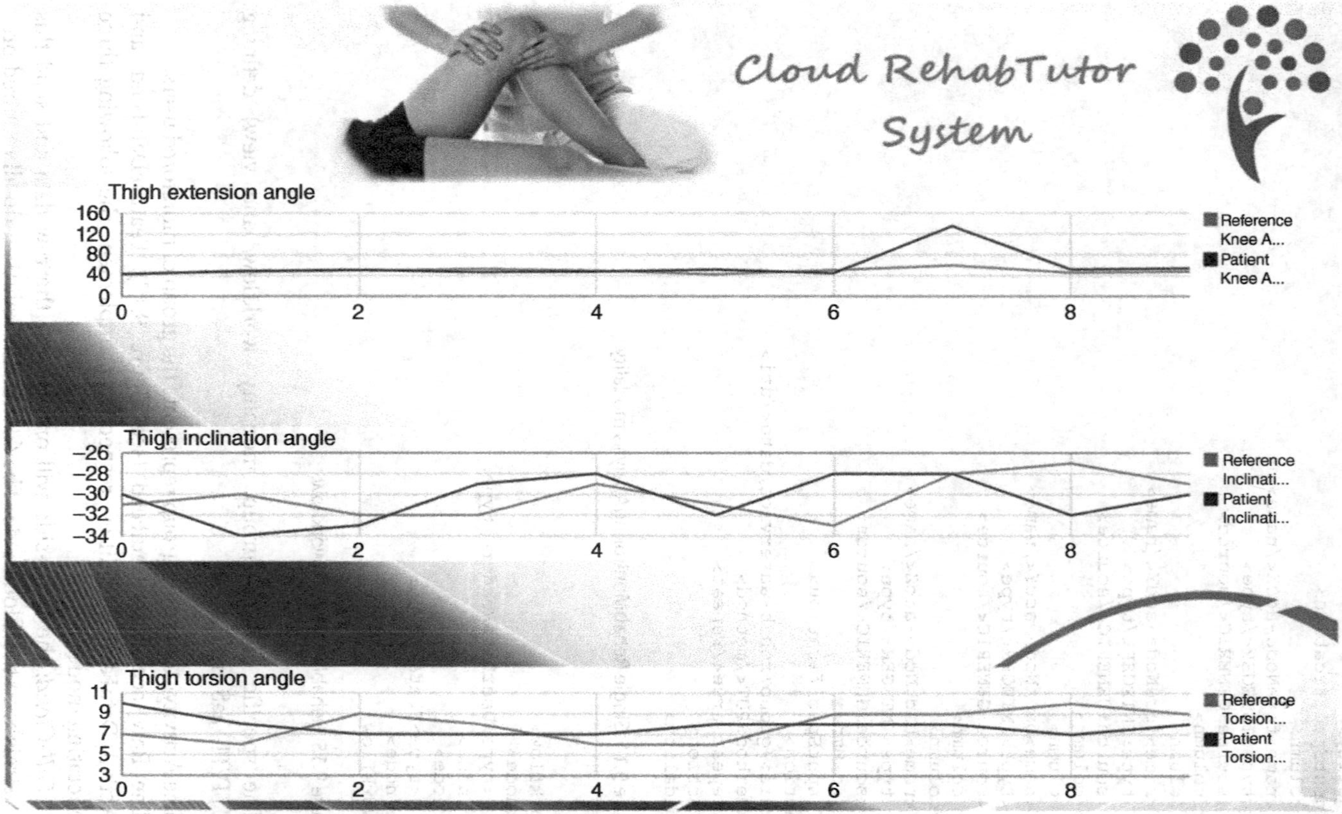

Figure 9.16 GUI view: knee rehabilitation.

```
<modality>
 <inputSpecification>
  <column>
   <name>acc_x_node1</name>
   <type>INTEGER</type>
   <source>GENERIC</source>
  </column>
  <column>
   <name>acc_y_node1</name>
   <type>INTEGER</type>
   <source>GENERIC</source>
  </column>
  <column>
   <name>acc_z_node1</name>
   <type>INTEGER</type>
   <source>GENERIC</source>
  </column>
  <column>
   <name>acc_x_node2</name>
   <type>INTEGER</type>
   <source>GENERIC</source>
  </column>
  <column>
   <name>acc_y_node2</name>
   <type>INTEGER</type>
   <source>GENERIC</source>
  </column>
  <column>
   <name>acc_z_node2</name>
   <type>INTEGER</type>
   <source>GENERIC</source>
  </column>
  <column>
   <name>geoLocation</name>
   <type>STRING</type>
   <source>GENERIC</source>
  </column>
  <column>
   <name>timestamp</name>
   <type>DOUBLE</type>
   <source>CLOCK</source>
  </column>
 </inputSpecification>
 <action>
  <uri>/group/fullCloud/data</uri>
  <method>PUT</method>
  <repeat>true</repeat>
  <trigger after="100" />
 </action>
</modality>
```

Figure 9.17 RawAccelerationDataFeed modality.

- *Mix-Cloud*: the Body-side will be responsible for raw data collection and feature extraction. These features will then be sent to the Cloud-side for classification.
- *Full-Local*: all processing will be done at the Body-side. Specifically, raw data collection, feature extraction, and feature classification. The

```
<modality>
 <init-action>
  <uri>/group/activity</uri>
  <method>GET</method>
 </init-action>
 <action>
  <uri>/engine/workflow/activity</uri>
  <method>POST</method>
  <parameter>
   <name>sourceUser</name>
   <reference xpath="//users/user" />
  </parameter>
  <parameter>
   <name>sourceGroup</name>
   <value>activity-recognition</value>
  </parameter>
  <repeat>false</repeat>
 </action>
 <outputSpecification>
  <column>
   <name>activityID</name>
   <type>INTEGER</type>
  </column>
  <view>/view/activities.xml</view>
 </outputSpecification>
</modality>
```

Figure 9.18 Single ActivityMonitoring Analysis modality.

```
<workflow>
 <node>
  <type>UserDataReader</type>
 </node>
 <node>
  <type>ACTSTATS</type>
  <!-- <parameter days="1" /> -->
 </node>
</workflow>
```

Figure 9.19 ActivityMonitoring workflow.

Cloud-side is therefore used only for long-term storage and graphical visualization of statistics.
- The Single ActivityMonitoring Analysis modality (see Figure 9.18) implements the activity recognition of a single subject.
- The ActivityMonitoring workflow (see Figure 9.19) models a three sequential node workflow able to (i) read body motion data collected by the reader node, (ii) extract the features from such data, and (iii) apply the activity classification algorithm. Such workflow is specifically activated when ACTIVITYaaS runs in Full-Cloud mode.
- The Activity View models the web-based graphical representation of the various activities being performed by the user. Currently it uses a simple pie chart and table for statistics visualization (see Figure 9.20).

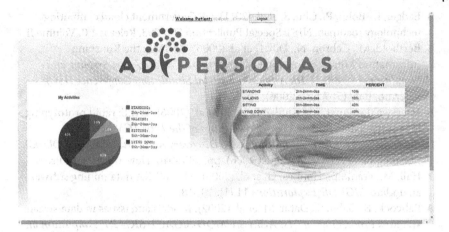

Figure 9.20 GUI view: activity statistics.

9.8 Summary

This chapter has provided an overview of the integration between wearable computing platforms (based on BSNs) and cloud computing, named cloud-based BSNs. We have first introduced the motivations and challenges for cloud-based BSNs. We have then introduced an implementation-neutral reference architecture for cloud-based BSNs. Furthermore, we have compared the related work against the analyzed requirements. Finally, the chapter has focused primarily on BodyCloud, a cloud-based BSN platform for the development of community BAN applications. A set of cutting-edge applications of BodyCloud have been also detailed to show the development effectiveness of BodyCloud.

References

1 Fortino, G., Giannantonio, R., Gravina, R. et al. (2013). Enabling effective programming and flexible management of efficient body sensor network applications. *IEEE Transactions on Human-Machine Systems* 43 (1): 115–133. doi: 10.1109/TSMCC.2012.2215852.

2 Foster, I. and Kesselman, C. eds. (2004). The grid 2 (second edition) blueprint for a new computing infrastructure. In: *The Morgan Kaufmann Series in Computer Architecture and Design*. Burlington: Morgan Kaufmann. doi: 10.1016/B978-155860933-4/50000-6.

3 Rimal, B.P., Choi, E., and Lumb, I. (2009). A taxonomy and survey of cloud computing systems. *Fifth International Joint Conference on INC, IMS and IDC, 2009. NCM'09*, Seoul, Korea (25–27 August 2009), pp. 44–51.

4 Badger, L., Bohn, R., Chu, S. et al. (2011). US Government cloud computing technology roadmap. NIST Special Publication 500-293, Release 1.0, Volume II.

5 Berthold, M., Cebron, N., Dill, F. et al. (2006). KNIME: the Konstanz Information Miner. *Proceedings of Workshop on Multi-Agent Systems and Simulation (MAS&S), 4th Annual Industrial Simulation Conference (ISC)*, Palermo, Italy (5–7 June 2006), pp. 58–61.

6 Mierswa, I., Wurst, M., Klinkenberg, R. et al. (2006). YALE: rapid prototyping for complex data mining tasks. *Proceedings of the 12th ACM SIGKDD International Conference on Knowledge Discovery and Data Mining (KDD'06)*. Philadelphia, PA (20–23 August 2006), pp. 935–940. New York: ACM Press.

7 Hall, M., Frank, E., Holmes, G. et al. (2009). The WEKA data mining software: an update. *SIGKDD Explorations* 11 (1): 10–18.

8 Babcock, B., Babu, S., Datar, M. et al. (2002). Models and issues in data stream systems. *Proceedings of 21st ACM SIGMOD-SIGACT-SIGART Symposium on Principles of Database Systems*, Santa Barbara, CA (21–24 May 2001), pp. 1–16. New York: ACM Press.

9 Golab, L. and Özsu, M. (2003). Issues in data stream management. *ACM SIGMOD Record* 32(2): 5–14.

10 Motwani, R., Widom, J., Arasu, A. et al. (2003). Query processing, resource management and approximation in a data stream management system. *Proceedings of International Conference on Innovative Data Systems Research (CIDR'03)*, Asilomar, CA (9–12 January 2003).

11 Aberer, K., Hauswirth, M. and Salehi, A. (2007). Infrastructure for data processing in large-scale interconnected sensor networks. *Proceedings of Int'l Conference on Mobile Data Management (MDM'07)*, Mannheim, Germany (7–11 May 2007).

12 Chandrasekaran, S., Cooper, O., Deshpande, A. et al. (2003). TelegraphCQ: continuous dataflow processing. *Proceedings of International Conference on Innovative Data Systems Research (CIDR'03)*, Asilomar, CA (9–12 January 2003).

13 Abadi, D., Carney, D., Çetintemel, U. et al. (2003). Aurora: a new model and architecture for data stream management. *The VLDB Journal* 12(2): 120–139.

14 Arvind, D., Arasu, A., Babcock, B. et al. (2003). STREAM: the Stanford stream data manager. *IEEE Data Engineering Bulletin* 26.

15 Delin, K. and Jackson, S. (2001). The sensor web: a new instrument concept. *Proceedings of SPIE Symposium on Integrated Optics*, San Jose, CA (20–26 January 2001).

16 Shen, C., Srisathapornphat, C., and Jaikaeo, C. (2001). Sensor information networking architecture and applications. *IEEE Wireless Communications* 8(4): 52–59.

17 Buonadonna, P., Gay, D., Hellerstein, J. et al. (2005). Task: sensor network in a box. *Proceedings of 2nd European Conference on Wireless Sensor Networks*, Istanbul, Turkey (31 January–2 February 2005), pp. 133–144.

18 Gay, D., Levis, P., von Behren, R. et al. (2003). The nesC language: a holistic approach to networked embedded systems. *SIGPLAN Not* 38(5): 1–11. doi:10.1145/780822.781133.

19 Malan, D., Fulford-Jones, T., Welsh, M., and Moulton, S. (2004). Codeblue: an
    ad hoc sensor network infrastructure for emergency medical care. *Proceedings
    of Internationall Workshop on Wearable and Implantable Body Sensor
    Networks*, London, UK (6–7 April 2004).
20 Kuryloski, P., Giani, A., Giannantonio, R. et al. (2009). DexterNet: an open
    platform for heterogeneous body sensor networks and its applications. *Sixth
    International Workshop on Wearable and Implantable Body Sensor Networks,
    2009. BSN 2009*, Berkeley, CA (3–5 June 2009), pp. 92, 97. doi: 10.1109/
    BSN.2009.31.
21 Bellifemine, F., Fortino, G., Giannantonio, R. et al. (2011). SPINE: a domain-
    specific framework for rapid prototyping of WBSN applications. *Software:
    Practice and Experience* 41 (3): 237–265. doi: 10.1002/spe.
22 Gravina, R., Guerrieri, A., Fortino, G. et al. (2008). Development of body
    sensor network applications using SPINE. *Proceedings of IEEE International
    Conference on Systems, Man, and Cybernetics (SMC 2008)*, Singapore (12–15
    October 2008).
23 Raveendranathan, N., Galzarano, S., Loseu, V. et al. (2012). From modeling to
    implementation of virtual sensors in body sensor networks. *IEEE Sensors
    Journal* 12 (3): 583–593.
24 Fortino, G., Guerrieri, A., Giannantonio, R., and Bellifemine, F. (2009).
    Platform-independent development of collaborative WBSN applications:
    SPINE2. *Proceedings of IEEE International Conference on Systems, Man, and
    Cybernetics (SMC 2009)*, San Antonio, TX (11–14 October 2009).
25 Fortino, G., Guerrieri, A., Giannantonio, R., and Bellifemine, F. (2009).
    SPINE2: developing BSN applications on heterogeneous sensor nodes.
    Proceedings of IEEE Symposium on Industrial Embedded Systems (SIES'09),
    special session on wireless health, Lausanne (8–10 July 2009).
26 Lombriser, C., Roggen, D., Stager, M., and Troster, G. (2007). Titan: a tiny
    task network for dynamically reconfigurable heterogeneous sensor
    networks. In *Kommunikation in Verteilten Systemen (KiVS)*. Berlin
    Heidelberg: Springer.
27 Dourish, P. (1995). The parting of the ways: divergence, data management and
    collaborative work. *Proceedings of 4th conference on European Conference on
    Computer-Supported Cooperative Work*, Stockholm, Sweden (10–14
    September 1995), p. 230.
28 Cuzzocrea, A., Fortino, G., and Rana, O.F. (2013). Managing data and processes
    in cloud-enabled large-scale sensor networks: state-of-the-art and future
    research directions, *13th IEEE/ACM International Symposium on Cluster
    (Cloud and Grid Computing)*, Delft, the Netherlands (13–16 May 2013).
29 Hanson, M.A., Powell, H., Barth, A.T. et al. (2009). Body area sensor networks:
    challenges and opportunities. *IEEE Computer* 42 (1): 58–65.
30 Le, T.T. and Moh, S. (2015). Interference mitigation schemes for wireless body
    area sensor networks: a comparative survey. *Sensors* 15: 13805–13838.
    doi:10.3390/s150613805.

31 Sha, K. and Shi, W. (2008). Consistency-driven data quality management of networked sensor systems. *Journal of Parallel and Distributed Computing* 68: 1207–1221.

32 Fortino, G., Pathan, M., and Di Fatta, G. (2012). BodyCloud: integration of cloud computing and body sensor networks. *IEEE International Conference and Workshops on Cloud Computing Technology and Science (CloudCom 2012)*, Taipei, Taiwan (3–6 December 2012).

33 Fortino, G., Di Fatta, G., Pathan, M., and Vasilakos, A.V. (2014). Cloud-assisted body area networks: state-of-the-art and future challenges. *Wireless Networks* 20 (7): 1925–1938.

34 Tan, C.C., Wang, H., Zhong, S., and Li, Q. (2008). Body sensor network security: an identity-based cryptography approach. *Proceedings of the First ACM Conference on Wireless Network Security (WiSec'08)*, Alexandria, VA (31 March–2 April 2008), pp. 148–153. New York: ACM Press.

35 Ming, L., Lou, W., and Ren, K. (2010). Data security and privacy in wireless body area networks. *IEEE Wireless Communications* 17 (1): 51–58. doi: 10.1109/MWC.2010.5416350.

36 Zhang, M. and Sawchuk, A.A. (2009). A customizable framework of body area sensor network for rehabilitation. *Second International Symposium on Applied Sciences in Biomedical and Communication Technologies, 2009. ISABEL 2009*, Bratislava, Slovak Republic (24–27 November 2009), pp. 1–6.

37 Klein, A. and Lehner, W. (2009). How to optimize the quality of sensor data streams. *Fourth International Multi-Conference on Computing in the Global Information Technology, 2009. ICCGI'09*, Cannes/La Bocca, France (23–29 August 2009), pp. 13–19.

38 Galzarano, S., Fortino, G., and Liotta, A. (2012). Embedded self-healing layer for detecting and recovering sensor faults in body sensor networks. *IEEE International Conference on Systems, Man and Cybernetics (SMC 2012)*, Seoul, South Korea (14–17 October 2012), pp. 2377–2382.

39 Chang, F., Dean, J., Ghemawat, S. et al. (2016). Bigtable: a distributed storage system for structured data. *8th USENIX Symposium on Operating Systems Design and Implementation (OSDI 2006)*, San Diego, CA (6–8 November 2006), pp. 205–218.

40 Calder, B., Wang, J., Ogus, A. et al. (2011). Windows Azure storage: a highly available cloud storage service with strong consistency. *23rd ACM Symposium on Operating Systems Principles (SOSP 2011)*, Cascais, Portugal (23–26 October 2011), pp. 143–157.

41 Holmes, G., Donkin, A., and Witten, I.H. (1994). Weka: a machine learning workbench. *Proceedings of the 2nd Australia and New Zealand Conference on Intelligent Information Systems*, Brisbane, Australia (29 November 1994–2 December 2 1994).

42 Guazzelli, A., Zeller, M., Chen, W., and Williams, G. (2009). PMML: an open standard for sharing models. *The R Journal* 1 (1): 60–65.

43 Kurschl, W. and Beer, W. (2009). Combining cloud computing and wireless sensor networks. *Proceedings of 11th Int'l Conf. on Information Integration and Web-based Applications & Services*, Kuala Lumpur, Malaysia (14–16 December 2009), pp. 512–518.

44 Chu, X. and Buyya, R. (2007). Service oriented sensor web. *Sensor Networks and Configuration*, pp. 51–74. Secaucus, NJ: Springer-Verlag New York, Inc.

45 Yuriyama, M. and Kushida, T. (2010). Sensor-cloud infrastructure-physical sensor management with virtualized sensors on cloud computing. *Proceedings of International Conference on Network-based Information Systems (NBiS'10)*, Takayama, Gifu, Japan (14–16 September 2010), pp. 1–8.

46 Di Stefano, S., Merlino, G., Puliafito, A. (2012). SAaaS: a framework for volunteer-based sensing clouds. *Parallel and Cloud Computing* 1 (2): 21–23.

47 Pandey, S., Voorsluys, W., Niu, S. et al. (2011). An autonomic cloud environment for hosting ECG data analysis services. *Future Generation Computer Systems* 28 (1): 147–154.

48 Lounis, A., Hadjidj, A., Bouabdallah, A., Challal, Y. (2012). "Secure and scalable cloud-based architecture for e-Health wireless sensor networks. *21st International Conference on Computer Communications and Networks (ICCCN), 2012*, Munich, Germany (30 July 2012–2 August 2012), pp. 1–7.

49 Wan, J., Zou, C., Ullah, S. et al. (2013). Cloud-enabled wireless body area networks for pervasive healthcare. *IEEE Network* 27 (5): 56–61.

50 Fortino, G., Gravina, R., Guerrieri, A., and Di Fatta, G. (2013). Engineering large-scale body area networks applications. *Proceedings of 8th Int'l Conference on Body Area Networks (BodyNets)*, Boston, MA (30 September–2 October 2013).

51 Fortino, G., Parisi, D., Pirrone, V., and Di Fatta, G. (2014). BodyCloud: a SaaS approach for community body sensor networks. *Future Generation Computer Systems* 35: 62–79.

52 Andreoli, A., Gravina, R., Giannantonio, R. et al. (2010). SPINE-HRV: a BSN-based toolkit for heart rate variability analysis in the time-domain. *Wearable and Autonomous Biomedical Devices and Systems for Smart Environment, ser. Lecture Notes in Electrical Engineering*, vol. 75, pp. 369–389. Berlin/Heidelberg: Springer.

53 Covello, R., Fortino, G., Gravina, R. et al. (2013). Novel method and real-time system for detecting the Cardiac Defense Response based on the ECG. *IEEE International Symposium on Medical Measurements and Applications (MeMeA 2013)*, Ottawa, Canada (4–5 May 2013).

54 Gravina, R., Fortino, G. (2016). Automatic methods for the detection of accelerative cardiac defense response. *IEEE Transactions on Affective Computing* 7 (3): 286–298.

55 Fortino, G. and Gravina, R. (2014). Rehab-aaService: a cloud-based motor rehabilitation digital assistant. *2nd ICTs for improving Patient Rehabilitation Research Techniques Workshop*, Oldenburg, Germany (20 May 2014).

56 Fortino, G., Gravina, R., and Russo, W. (2015). Activity-aaService: Cloud-assisted, BSN-based system for physical activity monitoring. *Proceedings of IEEE CSCWD 2015*, Calabria (6–8 May 2015).

57 Gravina, R., Ma, C., Pace, P. et al. (September 2016). Cloud-based activity-aaService cyberphysical framework for human activity monitoring in mobility. *Future Generation Computer Systems* 75: 158–171.

10

Development Methodology for BSN Systems

10.1 Introduction

Designing BSN systems is a complex task and formal methods should be adopted to obtain correct, efficient, and cost-effective solutions. The most common approach is *bottom-up*: hardware components are chosen "a priori," followed by the communication protocols, and finally, applications are programmed atop the identified underlying infrastructure. The opposite design approach is *top-down*: high-level application requirements, driving the design process, are mapped to application-level frameworks, i.e. a set of programming abstractions and libraries; protocol stacks and hardware platforms are defined subsequently.

This chapter describes a development methodology for BSN systems, based on the SPINE framework, that follows a hybrid hardware–software codesign approach inspired to the Platform-Based Design (PBD).

10.2 Background

PBD [1] has been originally introduced as a methodology for the design of traditional embedded systems and more recently for WSNs. This methodology defines the design as a sequence of steps that lead from the initial high-level system description down to the actual implementation. Each step is an iterative refinement process that translates a higher level description to a lower level one that is progressively closer to the final implementation. Each refinement step is obtained by *mapping* all the components of the higher level description with components (or composition of components) from a lower level description. The mapping results from solving a constrained optimization problem: the choice is a mapping that satisfies the higher level description constraints while optimizing according to a cost function defined by the designer. For each

Wearable Computing: From Modeling to Implementation of Wearable Systems Based on Body Sensor Networks, First Edition. Giancarlo Fortino, Raffaele Gravina, and Stefano Galzarano.
© 2018 John Wiley & Sons, Inc. Published 2018 by John Wiley & Son, Inc.

layer of abstraction, these components, along with a description of their interfaces and performance, are stored in a library, called *platform*. The higher the initial level of abstraction, the easier is formulating functionalities and constraints, but the more difficult is to reach to a high-quality translation due to the semantic gap between specification and implementation.

Each refinement step is performed with a hybrid approach, where application constraints are refined in a top-down fashion, architecture performance are abstracted in a bottom-up fashion, and a *meet-in-the-middle* phase decides the actual implementation as discussed above.

The formalization of the PBD methodology is based on the Agent Algebra [2], which represents a formal tool to describe the refinement process. The refinement is the expression of a *function* in terms of the elements of a platform.

Three domains of agents are used to describe the mapping process and performance evaluation: the first two represent, respectively, the platform and the function; the third, referred to as *common semantic domain* (CSD), is an intermediate domain to map functions onto platform instances. A platform, depicted on the right in Figure 10.1, corresponds to the *implementation search space*. The function, on the left in Figure 10.1, corresponds to the *specification domain*. The function and the platform meet in the CSD. This domain plays the role of a common refinement and is used to combine the properties of both the platform and the specification domain that are relevant to the mapping process. The function is mapped onto the CSD as depicted in Figure 10.2. A platform instance is projected onto the CSD by considering the agents that can be implemented with that particular instance. This projection, represented by the arrows that originate from the platform in Figure 10.2, may or may not have a greatest element. If it does, the greatest element represents the nondeterministic choice of the functions that are implementable by the instance.

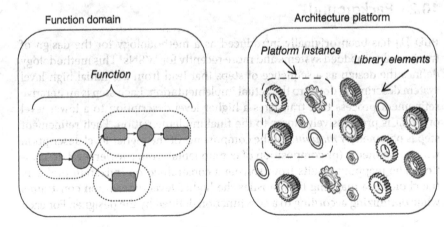

Figure 10.1 Architecture and function platforms.

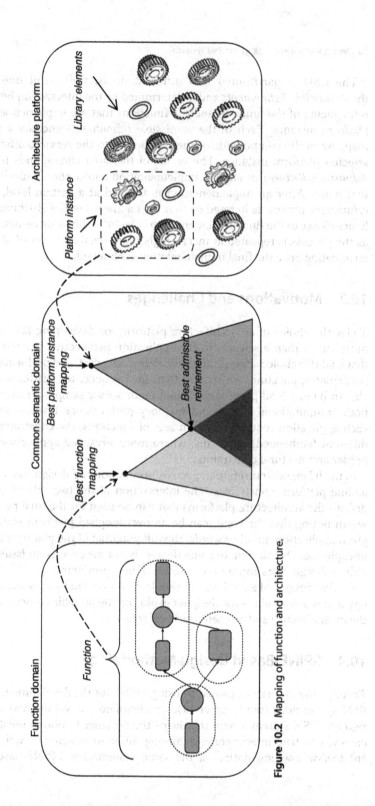

Figure 10.2 Mapping of function and architecture.

The CSD is partitioned into different areas: the useful one contains the *Admissible Refinements* and is determined by the intersection between the refinements of the function and the functions that are implementable by the platform instance. Each of the admissible refinements encodes a particular mapping of the components of the function onto the services offered by the selected platform instance. The vertex of this area corresponds to the *Best Admissible Refinement* and the implementation choice should ideally bring to that point. After an implementation is selected at a certain level, the same refinement process is iterated so to obtain a lower level of abstraction that is hence closer to the final implementation. The PBD shows its recursive nature, as the process is repeated at increasingly more detailed levels of abstraction, terminating once the final implementation is obtained.

10.3 Motivations and Challenges

Today the choice of an architecture platform for developing BSN systems is more an art than a science. In the application perspective, the requirements that lead this choice are typically wearability, size, cost, and performance. For a particular application, we require that, for instance, the platform should be able to handle (and preprocess) a minimum sensor sampling rate, which has both computational power and memory performance involvements. Since each application requires different sets of functions, the constraints identify different (embedded) platforms, where more advanced applications yield to harder architectural constraints.

In the IC manufacturers' perspective, production and design costs also imply adding platform constraints. The intersection of the two sets of constraints defines the architecture platforms that can be used for the final product. It is worth noting that the result can be an overdesigned platform instance for a given application; in other words, the full potential of the platform is partially unexploited. Overdesign, to some degree, is not necessarily an issue, as it can reduce design costs and time-to-market of new products.

So, the "design" of a BSN system should be supported by a formal methodology that is able to allow the designer exploring the possible choices looking for the most effective and efficient trade-off solution.

10.4 SPINE-Based Design Methodology

Through the experience gained by using SPINE for the development of several BSN applications (see Chapter 11), we identified a novel method to support rigorous BSN system design that helps the designer to obtain reliability, efficiency, and true interoperability among different systems as well as different hw/sw implementation of the same system. The SPINE-based Design

Methodology (SPINE-based DM) is inspired by the well-known PBD [1]. Here, however, the necessary platforms are opportunely semi-instantiated.

Specifically, according to the PBD, and in particular following the indication of Ref. [3], three layers of abstraction and corresponding platforms have been defined: the *Service Platform* at the application layer, the *Protocol Platform* to formalize communication protocols, and the *Implementation Platform* to describe hardware devices. Each design integrates an instance of these layers. Specifically, at each given refinement step, the design consists of a complete instance of the BSN system under development. We identified three main refinement steps: high level, detailed design, and implementation.

However, our approach differentiates from the standard PBD methodology because, with the intent of guiding the designer during the development of a SPINE-based efficient BSN system, some of the platforms we identified are semi-instantiated. Specifically:

- The Service Platform is bound to the high-level API provided by the SPINE Framework (see Chapter 3). Application requirements and functionalities can be mapped freely to the flexible SPINE API and services.
- The Implementation Platform includes many hardware. The designer has the opportunity to choose the most suitable one according to low-level system requirements. The Implementation platform is semi-instantiated too, as we assume, at the sensor-node level, the use of TinyOS-based architectures onto which the node-side of the SPINE Framework has been deployed, and, at the coordinator level, the use of Java- and Android-powered personal devices/computers that will be used as SPINE-based BSN coordinators.
- The Protocol Platform allows choosing two protocol stacks: Bluetooth and IEEE 802.15.4. This platform is the last to be instantiated as the choice often depends on the mapping made at the Implementation Platform (particularly on the radio standard available on the target devices).

10.4.1 A Pattern-Driven Application-Level Design

The application-level design of a SPINE-based BSN application can be guided by pattern-driven strategies. In the following, we describe two of such useful design patterns, both completely supported by SPINE:

- *Sensor Data Collection for Monitoring*: The simplest pattern supports the development of BSN systems for data collection from a set of wearable sensors into the coordinator which, in turn, visualize, store, and/or analyze such collected data. The pattern architectural schema is depicted in Figure 10.3a. Its main components are organized in two layers:

1) *Sensing*, in which data are collected from the sensor nodes.
2) *Monitoring*, in which data can be visualized, analyzed, and stored.

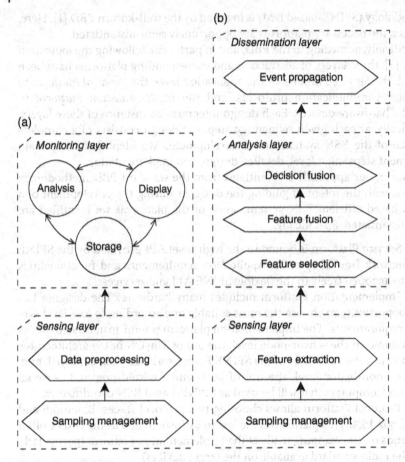

Figure 10.3 Pattern architectural schemas: (a) Sensor Data Collection for Monitoring; (b) Multisensor Data Fusion for Detection/Classification of Events.

Each layer can be implemented either at the sensor or coordinator level. At the Sensing layer, the sampling management component feeds the data preprocessing component with sensory data. At the monitoring layer, data can be stored by the data storing component, analyzed by the data analysis component, and graphically visualized by the data visualization component. It is worth noting that none of these components are required; each of them can be optionally included.

- *Multisensor Data Fusion for Detection/Classification of Events*: This pattern extends the previous by introducing the detection and/or classification of events of interest, such as accidental falls, physical activities, posture or gestures, mental states, and so on (see Chapter 11). Its architectural schema is depicted in Figure 10.3b. The main components are organized in three layers:

1) *Sensing*, defined as for the previous pattern.
2) *Analysis*, in which decisions are inferred from available sensory data.
3) *Dissemination*, in which extracted information is provided to end-user BSN applications.

Each layer can be implemented either at the sensor or coordinator level. At the Sensing layer, the sampling management component feeds the feature extraction component that, in turn, extracts features such as the maximum/minimum values, signal energy, or average value. At the Analysis layer, (i) the feature selection component contains algorithms for the selection of the most significant feature sets, (ii) the feature fusion component merges the different features together, and (iii) the decision fusion component, on the basis of the incoming features set, performs decisions such as classification of human postures or gestures (see also Section 11.3). Finally, at the Dissemination layer, the event propagation component forwards such decisions to (local and/or remote) application-level components.

10.4.2 System Parameters

According to the proposed method terminology, the main parameters affecting BSN-based applications can be classified as follows:

1) *Application-level* parameters: system accuracy, reliability, and responsiveness. Accuracy is application-specific and related to pattern recognition and event classification such as activity recognition or stress detection accuracy (see also Sections 11.2 and 11.3). Reliability is very relevant for life-critical applications (e.g. early detection of cardiac attacks, epilepsy attacks, and fall detection). A fuzzy definition of responsiveness is the ability of a system to provide the necessary feedback to the user within acceptable times; it is application-specific too, as it depends on the processing load required to perform the main operations, e.g. computation of flexion/rotation degrees in motor rehabilitation digital assistants (see also Section 11.5), or detection of a handshake in a handshake detection system (see also Section 11.4).
2) *Protocol-level* parameters: bandwidth and delay that depend on sensor sampling frequency, sensor- and application-specific generated data, and on communication protocols themselves. It is worth noting, however, that specific network synchronization requirements can be handled by the selected protocol (e.g. by using a TDMA technique), whereas more complex synchronization constraints must be handled at the Application level.

3) *Device-level* parameters: energy consumption, memory, and processing capabilities requirements. The energy consumption depends on duty cycle, sensor type and sampling frequency, radio usage, and application-specific signal processing. Memory (system and mass memory) requirements depend on (i) software platform tailoring (i.e. specific to our design method, for which TinyOS and SPINE components are needed), (ii) sampling frequency, (iii) buffering allocation parameters for sensor data storing and computation (e.g. buffer pool size, window, and shift size), and (iv) on application-specific signal filtering and data processing. Computing power is mostly determined by application-specific signal processing.

10.4.3 Process Schema

The SPINE-based Platform Design [4] process schema is depicted in Figure 10.4. The process is iterative and is composed of the following steps (carried out by Modeler, Designer, and Developer roles):

- *Requirements Analysis (RA)*: it produces a set of functional and nonfunctional requirements driving the design flow.
- *High-Level Design (HLD)*: it produces a high-level design of the BSN system on the basis of the identified requirements. In our methodology, a HLD is an instance of the SPINE framework integrated with selected protocols, sensors, and platforms.
- *Performance Estimation of HLD*: it produces estimation measurements of the HLD performance by using available analytical/simulation methods. Although the results cannot be detailed at this refinement level, they can still provide insights on the feasibility (or convenience) to translate the available HLD into a DD. If the requirements are not satisfied, the process must step back to the HLD step.
- *Detailed Design (DD)*: it produces the detailed design of the available HLD instance. The HLD is refined at each of the three layers of SPINE-based DM by following the pattern-driven design described previously.
- *Performance Estimation of DD*: it provides analysis of the DD by testing or estimating the DD performance through analytical and/or simulation methods and also by mapping selected DD components onto the device level for testing. The obtained results are more accurate than the Performance Estimation HLD output; they provide fine-grained indications on the feasibility of obtaining an effective and efficient implementation of the DD instance. If the requirements are not satisfied, the process must go back to the DD or even the HLD step.
- *Implementation*: it produces an implementation of the DD output; the BSN system can be finally deployed, executed, and tested.
- *Deployment*: it defines deployment details of the BSN system.

Figure 10.4 SPINE-based Platform Design process schema.

- *System Performance Evaluation*: it provides detailed test cases of the BSN system and detailed performance measurements are extracted for its validation. The result of this analysis provides a full-fledged test of the whole system. If the requirements are not satisfied, the process must go back to the DD or even the HLD step.

10.5 Summary

This chapter has introduced a specialization of the PBD methodology for system-level design of BSN applications. First, the PBD approach has been briefly described. Then, a PBD methodology, previously proposed for the design of WSN systems, has been specialized for the more specific BSN domain. Finally, the methodology has been concretely shown in relation to the SPINE framework.

References

1 Keutzer, K., Newton, A.-R., Rabaey, J.-M., and Sangiovanni-Vincentelli, A. (2000). System-level design: orthogonalization of concerns and platform-based design. *IEEE Transactions on Computer-Aided Design of Integrated Circuits and Systems* 9 (12): 1523–1543.
2 Passerone, R. (2004). Semantic foundations for heterogeneous systems. PhD thesis. University of California.
3 Bonivento, A. (2007). Platform based design for wireless sensor networks. PhD thesis. University of California.
4 Fortino, G., Giannantonio, R., Gravina, R. et al. (2013). Enabling effective programming and flexible management of efficient body sensor network applications. *IEEE Transactions on Human-Machine Systems* 43 (1): 115–133. doi: 10.1109/TSMCC.2012.2215852.

11

SPINE-Based Body Sensor Network Applications

11.1 Introduction

The worldwide trend of increasing average life expectancy and a more profound awareness of the importance of taking actions at different levels to keep a good health status are forcing the health system to significant renovation. Enabling technologies in this context are the current powerful personal mobile devices, such as smartphones and tablets, the body sensor networks (BSNs), i.e. wearable sensor units (smart watches, glasses, and wristbands) that are often able to monitor several health parameters, and the cloud computing infrastructures. The result is a great opportunity of providing very diverse and personalized smart-Health services that could be accessible to anyone, anywhere, and anytime.

11.2 Background

This chapter emphasizes how the SPINE framework is actually able to support the development of heterogeneous health-care applications based on reusable subsystems. Indeed, one of the main goal of SPINE (see Chapter 3) is to provide a flexible architecture that can support a variety of practical applications without the need for costly redeployment of the code running on sensor nodes. This chapter therefore introduces some interesting research BSN systems that have been developed atop SPINE. Furthermore, each of the described applications improved the current state-of-the-art, as described in the following sections.

11.3 Physical Activity Recognition

Physical activities play a fundamental role in human well-being; however, although people are now fully aware of their importance, they still need regular motivational feedback to maintain an active life style. Thus, the automatic

Wearable Computing: From Modeling to Implementation of Wearable Systems Based on Body Sensor Networks, First Edition. Giancarlo Fortino, Raffaele Gravina, and Stefano Galzarano.
© 2018 John Wiley & Sons, Inc. Published 2018 by John Wiley & Son, Inc.

recognition of activities and postures is the first step for providing the right feedback. To this extent, physical activity recognition is a basilar block of many wellness and smart medical applications. In addition, many human-centric context-aware real-life applications need to assess user activities as they often heavily contribute at determining the context itself.

11.3.1 Related Work

Human activity recognition has attracted tremendous interest and the topic has been studied under very diversified point of views and the related issues addressed through different approaches in terms of types of sensory signals and recognition strategies. Research in physical activity monitoring is currently focused to support elderly people and patients with chronic diseases.

One of the most relevant and cited related work is by Bao and Intille [1]. In this study, several supervised learning algorithms are used and evaluated to detect physical activities using accelerometer data gathered from sensor nodes placed on different body locations. Acceleration data was collected from 20 subjects without researcher supervision or observation.

The authors in Ref. [2] address the very interesting aspect of comparing the activity classification accuracy by varying the number and the location of sensor nodes on the human body.

In Ref. [3], the authors propose an activity recognition system based on a single motion sensor node worn at the waist. Three axial acceleration signals are processed to extract significant features such as mean, standard deviation, energy, and correlation. A number of classifier algorithms (decision trees, K-nearest neighbors, SVM, and Naive Bayes) have been evaluated to assess their performance in terms of recognition accuracy. Furthermore, meta-level classifiers based on different approaches (voting, stacking, and cascading) have been taken into account too.

In Ref. [4], the focus is on the importance of designing power-aware recognition algorithms as they are implemented on power-constrained wearable devices. The authors investigate the benefits of dynamic sensor selection to achieve the best trade-off among power consumption and activity recognition accuracy and propose an activity recognition method that is associated to an underlying runtime sensor selection scheme.

In the last year, thanks to the tremendous improvements of commercial smartphones, not only in terms of computational and storage capabilities, but in particular of sensing opportunities, many research projects and commercial applications are highlighting the convenience of developing physical activity monitoring systems (as well as more generic smart-Health applications) solely supported by smartphones, so to significantly improve user acceptance and reduce economic costs. For instance, Ref. [5] presents a daily activity monitoring system designed for elderly people based on the smartphone accelerometer. The authors take into account the energy limitation and propose a power-aware approach as an adaptation of the standard Support Vector Machine (SVM). In Ref. [6], along with the accelerometer, the gyroscope and the magnetometer (available in many current smartphones) are also used to detect physical activities. The authors notably evaluate the effect on classification performances of smartphone position and orientation on the body.

Excellent review works on human activity recognition have been published too. In Ref. [7], the authors provided a review of the most relevant approaches and methodologies related to sensor-based activity monitoring, modeling, and recognition; advantages and weakness are discussed for each approach. An extensive survey [8] covers the state-of-the-art in human activity recognition, specifically based on wearable sensors. The authors propose a two-level taxonomy associated with the learning approach (supervised or semisupervised) and the response time (off-line or online).

11.3.2   A SPINE-Based Activity Recognition System

The human activity monitoring system presented here takes advantage of the past work aiming at finding the best trade-off among accuracy, wearability, power requirements, and programming complexity. It is able to recognize postures (lying down, sitting, and standing still) and a few movements (walking and jumping); furthermore, it also includes a simple yet effective fall detection module that uses the activity classification to determine if a person is unable to stand up after the fall.

The system uses two wireless wearable nodes based on the Shimmer2R [9] platform, which includes a 3-axis accelerometer and an Android-based personal mobile device (e.g. a smartphone or tablet) that acts as a coordinator. The end-user application (see Figure 11.1) runs on Android and is programmed atop the SPINE-Android framework. The sensor nodes and the coordinator communicate over Bluetooth.

The activity recognition system uses a classifier algorithm that takes accelerometer data gathered by the wearable units, placed on the waist and on the thigh of the assisted living, and recognizes gestures and activities defined during an off-line training step. Among the most popular classification algorithms used in the literature to this purpose, a K-Nearest Neighbor [10] (k-NN)-based classifier has been modeled.

The proposed system includes a default training-set to use the application without customization. However, a graphical wizard can be optionally used to improve recognition accuracy by creating a user-specific training-set. The most significant features for discriminating the different activities will be eventually computed online by the sensor devices but are initially identified with an off-line sequential forward floating selection (SFFS) [11] algorithm.

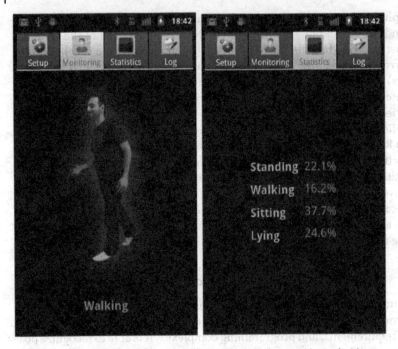

Figure 11.1 Two screenshots of the developed activity recognition Android app.

The k-NN classifier requires the selection of two different parameters: the value of K and the metric distance. However, if the feature selection process is performed accurately, the result will lead to activities' clusters that are internally very dense, and well separated among each other. This is particularly true on the specific set of activities targeted by the system. Therefore, the classifier accuracy is significantly influenced by its parameter values, which have been selected as follows, mainly to reduce classification execution time:

- $K = 1$
- Metric distance: Manhattan

The most significant feature set obtained with the SFFS algorithm is the following:

- *Waist node*: (i) average value on the accelerometer axes XYZ, (ii) minimum value on the accelerometer axis X, and (iii) maximum value on the accelerometer axis X.
- *Thigh node*: minimum value on the accelerometer axis X.

As aforementioned, the proposed system also integrates a fall detection functionality, whose underlying algorithm is distributed since it is partially

Table 11.1 Posture/movement recognition accuracy.

Sitting	Standing	Lying down	Walking
96%	92%	98%	94%

running on the waist node and partially on the mobile coordinator. Specifically, the algorithm computes in real time on the waist node the total energy (i.e. the square root of the sum of squares) over the three accelerometer axes. The instantaneous total energy value is compared against an empirically estimated threshold and if this is exceeded, the node triggers a "potential-fall" alarm message back to the coordinator. If such a preliminary alarm is received, the portion of the algorithm running on the coordinator starts monitoring the user postures for a certain period. If the user is detected as "lying down", an emergency message is reported to relatives and/or medical personnel via several channels (SMS and automated voice call to an emergency list of numbers, and even Facebook and Twitter posts). In particular, we differentiate two types of alarms: *yellow* if the user is able to stand up shortly after the fall, *red* if, after a few minutes, he/she is still lying down.

The classification accuracy performance achieved by the system, reported per each activity in Table 11.1, reaches an overall average score of 97%. The fall detection algorithm, instead, in a semicontrolled laboratory setting obtained an average accuracy of 90%, with a very low percentage of false alarms (less than 1%).

11.4 Step Counter

Human footstep detection refers to the automatic determination of the time moment at which steps occur. It is the basic block for the realization of step counters, also known as pedometers, which can be used to roughly assess in real-time human activity levels, which in turn is one of the major goal of wellness applications. Step counters have also been used to assess elderly mobility and to improve physical activity in youth to reduce the risk of obesity.

11.4.1 Related Work

Step detection has been broadly addressed and many different methodological and technological approaches have been proposed in the literature. A comprehensive review on this topic is out of the present scope and the interested reader can refer to Refs. [12, 13] for a deeper analysis. In the following, only few significant works, addressing the human step detection by means of wearable devices and accelerometer sensors, will be introduced.

In Ref. [14], a method for online step detection using an embedded device based on the IMote2 platform and equipped with a 3-axis accelerometer is presented. The device, which must be worn on the hip, samples the accelerometer at a frequency of 512 Hz. The raw acceleration signals are initially used to extract a cross-axial magnitude signal which is, in turn, smoothed with a low-pass filter. Then, the obtained signal is further processed to obtain its first derivative signal. Finally, threshold-based peak detection is performed.

In Ref. [15], a system specifically designed to assess the number of steps taken during running is presented. It is based on the Nokia Wrist–Attached Sensor Platform equipped with a 3-axis accelerometer. In this work, the acceleration signals are processed with a high-pass filter with the intent of removing the gravity component. The three high-pass filtered signals are then combined to generate a unique signal by taking the 1-norm, obtained by summing up the three axes' corresponding absolute sample values. Then, threshold-based peak detection is performed. It is worth noting that in this work the threshold is dynamically adapted. The overall system performance reaches a 30% underestimation of the actual number of steps taken while running.

In Ref. [16], a pedometer based on a custom prototype device using the 3-axis accelerometer ADXL330 connected to the 8-bit MPC82G516 microcontroller is presented. The device is intended to be worn on the waist or in the pocket. The raw acceleration signals are first smoothed with a Hamming filter. The x, y, and z acceleration vectors are used to evaluate the initial spatial orientation of the device so as to allow for an arbitrary placement of the device itself (particularly useful if it would be placed in the pocket). The filtered x, y, and z signals are also used to generate the acceleration signal in the direction of gravity. This latter signal is compared against a fixed, empirically evaluated, footstep threshold. The system has been evaluated in a laboratory setting on five subjects showing an average detection accuracy of about 90%.

11.4.2 A SPINE-Based Step Counter

This section describes an innovative step-counter algorithm that has been integrated into the previously described SPINE-based activity recognition application as an optionally activated functionality. To provide an original contribution and improvement to the state-of-the-art, we identified a number of key design requirements:

- Use of accelerometer data.
- Low sampling rate.
- Energy- and computation-efficient design to support embedded implementations.
- Use of a single sensor node, placed on the waist (below the navel).
- General-purpose algorithm, to be used by healthy people as well as elderly and/or people with disabilities.
- No need for "ad-personam" calibration.
- High average accuracy (robustness).

Several real walk data on different subjects have been collected and studied before starting the algorithm design. The subjects were asked to walk naturally and to increase/decrease the walking speed occasionally. In particular, a single 3-axis accelerometer sensor node was placed on the waist while recording. The sensor has been sampled at 40 Hz. To simplify the development, debugging, and evaluation, the algorithm has been initially programmed in Matlab. Only integer-math computations were used, thus allowing for a more straightforward embedded implementation (as the target embedded platform is based on a microcontroller with no hardware support for floating point operations).

It is worth noting that the frontal acceleration of the waist (i.e. parallel to the ground) presents a signal roughly sinusoidal while walking. The basic idea is, therefore, to detect steps by identifying the decreasing segment (falling edge), which corresponds to the last fraction of a step movement.

Furthermore, it is clear that a human step is characterized by time constraints (physically, it cannot be "too" fast or "too" slow). However, walk patterns change from people to people and even for the same person it might change from time to time; hence, the amplitude of the acquired signal can vary significantly.

To simplify recognition of the step pattern, the raw frontal acceleration is first processed with a smoothing filter, which removes the high-frequency components. Then, the algorithm looks for local maximums. When a local maximum is found, it looks for a local minimum. After the local minimum is also found, the candidate segment is naturally identified.

Two features are then extracted and used to determine whether the candidate belongs to an actual step or to different body movements. Specifically, the candidate is classified as step (i) if they have an acceleration drop within a certain range (specified by a "tolerance" parameter around a threshold) and (ii) if the time elapsed is within a certain interval. The preprocessing is a 9-point windowed smoothing filter, which uses Gaussian kernels. Because they are applied to a digital signal, the sum of the kernels must be 1. Furthermore, because the algorithm works on integer-math, they are scaled so that decimal factors are removed.

The threshold is coarsely initialized, but it is automatically adapted while steps are recognized. In particular, it is continuously updated with the average of the last 10 acceleration drops that are classified as steps. This is very useful to avoid custom training or a setup phase before the step-counter functionality could work properly and accurately. Finally, to reduce "false positive" recognitions, e.g. due to sudden shocks or slow tilts of the sensor, the time elapsed between the local max and min (which is simply determined as the product between the number of samples of the segment and the sampling time) must be longer than the "minimum step time" and shorter than the "maximum step time." Both values have been determined empirically from the available observations.

The proposed algorithm has been initially evaluated on the computer and finally implemented on a wireless sensor node running SPINE. For this application, the node-side of SPINE has been extended with the proposed algorithm. Every time the node detects a step, it communicates to its coordinator the total number of steps taken so far, in order to avoid miscounting due to lost packets. On the SPINE coordinator, very minor additions have been made to the core framework, and a simple graphical gadget has been added to show in real time the number of steps being taken.

11.5 Emotion Recognition

Emotions play a fundamental and basilar role in daily life of each person, both at the individual and social level. The need and importance of automatic emotion recognition is growing along with the increasing popularity of human–computer interface (HCI) systems. Today, in fact, new forms of human-centric interaction with digital media and devices have a disruptive potential of revolutionizing many aspects of virtual and real life. Furthermore, automatic emotion recognition could provide helpful medical information and indices for the prevention or early detection of many psychophysiological disorders.

Among the many human emotions, being able to automatically recognize stress and fear, thus, becomes very useful, as it will be described in the following two sections.

11.5.1 Stress Detection

The Heart Rate Variability (HRV) is based on the analysis of the R-peak to R-peak intervals (RR-intervals – RRi) of the electrocardiogram (ECG) signal in the time and/or frequency domains. In recent years, the importance of the HRV for detecting mental and emotional states is being recognized by physician and psychologists, specifically for the sake of identifying stress and anxiety.

11.5.1.1 Related Work

Past medical studies have showed that patients with anxiety, phobias, and stress disorders consistently present lower HRV. It is worth noting that this relationship exists independently of gender, age, heart and respiratory rate, trait anxiety, or blood pressure.

Monitoring the mental stress is particularly important because studies showed that long-term exposure to stress is a risk factor for cardiovascular diseases [17, 18]. Many industry research projects focus on HRV, looking for connections with related heart diseases. An interesting research [19] actually proves the existence of a relation between time-domain HRV parameters and stressful car driving situation.

In Ref. [20], the authors present an activity-aware mental stress detection approach using ECG, GSR, and accelerometer data. Specifically, the work is focused on sitting, standing, and walking.

In Ref. [21], an interesting application of the stress detection to biometric security is proposed. Furthermore, the work reviews several methods for stress detection, to assess which one is most suitable for implementation in future biometric devices.

There are also a few commercial products for mental stress assessment. For instance, StressEraser [22] provides a biofeedback of the stress level looking for the breathing pattern that maximizes Respiratory Sinus Arrhythmia. Stress Monitor [23] is another system designed for stress monitoring while working. It is composed of a USB ear-clip device to be connected to a PC and a desktop application for real-time and historical reports. Finally, the emWave Personal Stress Reliever [24] is a handheld device with audio and LED feedbacks to monitor the stress level of the user. It is worth noting that none of these commercial products are suitable for continuous monitoring as they must be handled in hand to work or they cannot operate standalone.

11.5.1.2 SPINE-HRV: A Wearable System for Real-Time Stress Detection

In this section, we present a wearable system programmed atop SPINE (opportunely extended with a custom-defined processing function) that uses time-domain HRV analysis to detect mental stress [25]. It is designed for continuous noninvasive use and consists of a wearable cardiac sensor node (we have two alternative implementations, one with a Shimmer2R node equipped with the add-on ECG sensor board and the other with a Polar Electro [26] ECG wireless chest band), which extracts the RRi from the full ECG signal. The RRi are then processed using the SPINE framework with an application running on the coordinator (see Figure 11.2).

In particular, the system extracts common parameters known in the medical literature to perform an HRV analysis applied for continuous noninvasive mental stress detection of people during everyday activities.

The stress detection is computed at regular intervals (tunable from 10 to 60 min). Our approach is based on a time-domain analysis, which is sufficiently accurate to recognize the stress condition as shown in Ref. [27]. In particular, the analysis involves the computation of four significant indices:

$$\overline{RR}_j = \frac{1}{15}\sum_{j=1}^{15} RR_j$$

$$SDNN = \sqrt{\frac{1}{N-1}\sum_{j=1}^{N}\left(RR_j - \overline{RR}\right)^2}$$

$$RMSSD = \sqrt{\frac{1}{N-1}\sum_{j=1}^{N-1}\left(RR_{j+1} - RR_j\right)^2}$$

$$pNN50 = \frac{NN50}{N-1}\times 100$$

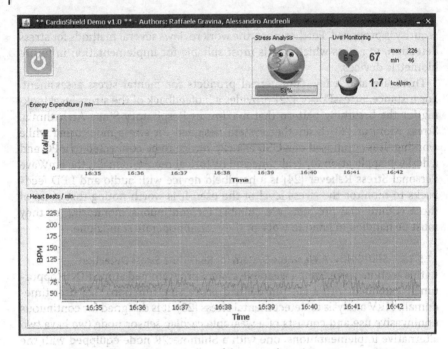

Figure 11.2 The main monitoring window of the stress detection system.

RR$_j$ denotes the value of the jth RR-interval and N is the total number of successive intervals. $\overline{RR_j}$ is therefore the average value of 15 consecutive RRi. SDNN, the standard deviation of RR$_j$, is the primary measure used to quantify HRV changes, since SDNN reflects all the cyclic components responsible for variability in the period of recording. RMSSD is the root mean square of successive differences. Finally, pNN50 is the ratio derived by dividing NN50 by the total number of RR$_j$, where NN50 represents the number of successive intervals differing by more than 50 ms.

The proposed system aims at detecting whether the monitored person is mentally stressed. This decision problem has been solved with a threshold-based approach.

Table 11.2 reports the threshold values extracted from the results found in Ref. [27]. The RR signal is recorded for a certain time window, at the end of which the algorithm computes the features, and, if at least three out of the four of them exceed the values reported in Table 11.2, the persons' mental state is classified as "stressed." In synthesis, the novelty in the proposed system is that the ability of performing online emotional stress detection rather than through off-line analysis.

Table 11.2 Stress threshold for HRV parameters.

Feature	Threshold	Unit
HR	>85	$1\ min^{-1}$
pNN50	<7	%
SDNN	<55	ms
RMSSD	<45	ms

11.5.2 Fear Detection

Fear is the physiological reaction in response to a danger or threat. Among the other psychophysiological reactions preceding the emotion of fear, there is a specific event that can be observed in the cardiac activity known as Cardiac Defense Response (CDR) [28, 29]. This response is the first of an internal process sequence preparing a reaction to threats priming either for fighting or fleeing (this is known as "fight-or-flight") [30]. In particular, right after a sudden situation perceived dangerous by the brain, the first basic reaction is the CDR activation. Then, if the stimulus is eventually classified as not actually dangerous, the organism goes back to a normal state and the heart rate (HR) stabilizes, otherwise a sense of fear will start to be perceived. Thus, the CDR has a protective and defensive role; nevertheless, if triggered too often and/or irrationally, it may represent a health risk and in the long term it could lead to several psychological disorders such as mental stress, phobia, anxiety, and depression [31]. Thus, being able to identify automatically the CDR activation is relevant, clinicians could be in fact aided by a valuable tool for studying the psychological state of the subject.

The ECG is being studied for emotion recognition and stress detection [25] as it has demonstrated the influence that psychological states due to emotions and other external conditions have on the ECG signal.

11.5.2.1 Related Work

The literature on the specific problem of automatic recognition of the human fear emotion is extremely limited. Mostly, past studies have investigated the broader problem of emotion recognition [32, 33] with controversial results. Some more related works focus instead on the arousal monitoring [34–36]. Arousal is a psychophysiological state of being awake or reactive to stimuli and plays a central role for motivating the fight-or-flight response, which in turn often precedes the emotion of fear.

11.5.2.2 A SPINE-Based Startle Reflex Detection System
This section introduces a SPINE-based mobile system that recognizes in real time basic emotional responses and in particular the CDR, which is triggered before the fear emotion itself [37, 38]. To the best of our knowledge, this is the first work aiming at recognizing automatically and in real time this physiological mechanism.

For the sake of realizing a portable noninvasive system, there are clear advantages on using the ECG, as technologies based on lightweight wearable cardiac sensors can be used. In particular, detecting the CDR requires the extraction of the RRi, and consequently the HR, from the full ECG trace.

We proposed an algorithm for the detection of the QRS complex (i.e. the heartbeat) inside the ECG signal using a dynamically adapted threshold-based approach. The algorithm looks for peaks in the ECG that are compared against an automatically estimated threshold; those exceeding the threshold are labeled as heartbeats and time stamped, hence leading to the RRi series, which are the input to the actual CDR detection algorithm. The proposed QRS detection algorithm runs on the personal mobile device and is part of the mobile application running atop the SPINE-Android coordinator.

In Figure 11.3, the schematic block diagram of the proposed adaptive QRS detection algorithm is shown. The algorithm consists of three main processing phases: a moving average-based high-pass filtering (HPF), a nonlinear low-pass filtering (LPF), and a decision-making block [39]. More specifically:

1) First, an ECG recording is processed by the linear HPF to amplify the QRS complex, while suppressing the undesired waveforms (e.g. P or T waves) and the baseline wander. This step consists of a 5-point moving average filter whose output is subtracted, point-by-point, from the delayed input sample so that the entire system becomes an FIR HPF with linear phase.
2) Then, the linear HPF output is then processed by a full-wave rectification and nonlinear amplification followed by a sliding-window summation, thus resulting in an envelope-like feature waveform. These operations (a nonlinear

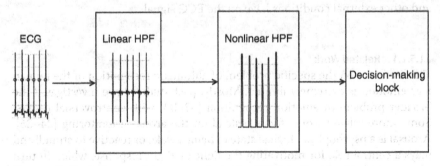

Figure 11.3 Block diagram of the proposed adaptive QRS detection algorithm.

LPF process) aim at smoothing down the high-frequency, low-amplitude artifacts while leaving the QRS waveform intact.

3) Finally, an adaptive threshold is applied to the feature waveform to complete the QRS complex detection.

To detect the CDR, we proposed an algorithm based on the idea of detecting changes in signal stationary. The underlying ratio is that physiological signals, including the ECG and its derived RR signal, are highly stationary. Formally, a signal is stationary if the mean and standard deviation of the signal do not change during signal acquisition. In ECG and RR signals, in particular, nonstationary events are due to several factors (e.g. changes in posture and respiration patterns).

Our intuition suggests that physiological changes and, more specifically, the effects of the CDR associated with responses to basic emotions such as the fear can also introduce nonstationary events in the ECG and consequently in the RR signal [28–31].

Thus, sudden changes in HR regulation due the CDR can be detected by looking at the nonstationary transitions from the normal HR regulation. The CDR algorithm adopts the cross-correlation integral method to quantify the amount of stationary in the given RR signal [40]. It provides the probability that a particular signal is stationary: a value close to 1 indicates a stationary signal; conversely, a value closer to 0 refers to highly nonstationary signals. We propose to compute the cross-correlation integral in a moving-window fashion (10% of the signal length). This allows for the detection of transitions in nonstationary in the RRi signal by running the CDR detection algorithm as a function of time. Finally, the cross-correlation integral samples are converted to percentages; this feature is referred as *nonstationary index* (NSI).

The CDR algorithm has been validated on 40 subjects, evaluating the NSI to establish the occurrence of the CDR. Specifically, a change pattern is classified as a CDR event if a reduction in the NSI is less than or equal to 80%. This specific NSI threshold was empirically estimated by direct observation of the data from all 40 subjects. The proposed system includes original contributions:

- It detects patterns in the HR signal, that is, it detects if the signal presents nonstationary transitions as they indicate changes in regulation of the HR signal.
- In contrast to related work in the psychological literature [28, 29], the CDR activation is detected in real time.
- By analyzing the CDR detection algorithm results, it is possible to locate the CDR event in the RR signal.

Figure 11.4 shows a portion of a real RR signal (top) and the corresponding NSI (bottom). The plot shows that a change in signal stationary can be observed when the subject experienced the CDR triggered by the external stimulus

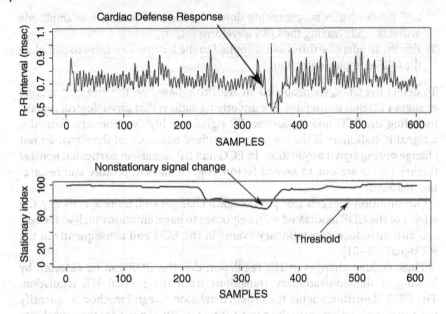

Figure 11.4 The proposed CDR detection algorithm applied to RRi series.

prompted to the subject during our experimental protocol setup. In particular, it is worth noting the NSI exceeding the 80% threshold.

The CDR detection algorithm was implemented with the R scripting language for the availability of mathematical and statistical libraries useful for the algorithm.

In addition, we realized a mobile application (see Figure 11.5), for devices running the Android OS, that is able to monitor the cardiac activity and, in particular, for detecting the CDR mechanism activation.

This system is implemented atop SPINE and uses a Shimmer2R node, equipped with the ECG sensor board, placed on the chest with a dedicated elastic band. The Android application uses the "Rserve" [41] library to communicate with an R server responsible for the remote execution of the CDR algorithm. Furthermore, the application also displays the current BPM value, the full ECG signal, the RRi series, and the historical HR chart.

11.6 Handshake Detection

The handshake is a basic gesture in many cultures. It introduces many formal and informal social interactions such as exchanging greetings, offering congratulations, or finalizing a deal. Thus, automatic handshake detection could enable

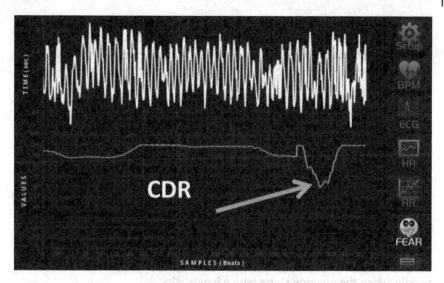

Figure 11.5 A screenshot of the developed CDR detection mobile application.

several pervasive computing scenarios; specifically, different types of information can be exchanged and processed among the handshaking people, for instance based on the physical/logical contexts and on mutual acquaintance.

11.6.1 Related Work

Very few research works on automatic handshake detection have been published so far. The iBand [42] is probably the first system for information exchange specifically based on handshake detection. It is based on wearable wrist devices equipped with accelerometers and infrared (IR) transceivers. Specifically, the handshake is detected via a synchronized combination of IR alignment and an up-and-down motion on the two devices worn by each meeting person. IR transmission is enabled when the user's hand/wrist are in a pre-calibrated handshaking orientation. The pre-calibration cannot be customized, thus leading, according to users participating in an experimentation of the system, to not always accurate behavior and unnatural gestures to let the iBand detect the handshake. Furthermore, a quantitative performance analysis of the system is not presented.

The Smart-Its Friends [43] provides smart electronic devices that communicate when they are within the communication range of each other and experience similar sensor readings. Although the proposed approach is more general, it could be applied in the context of handshake detection: handheld smart devices equipped with accelerometers (e.g. smartphones and augmented wrist watches) can be exploited to recognize common shaking patterns between people when they are in proximity, even if this would be an indirect way of detecting handshakes as Smart-Its Friends is not focused on the handshake gesture but only on generic interaction among smart objects close to each other.

11.6.2 A SPINE-Based Handshake Detection System

To overcome the limitations of the aforementioned works, a further interesting application developed with SPINE, called E-Shake (see Figure 11.6), has been proposed [44]. E-Shake is a Collaborative BSN-based system for the detection of emotions between people as they shake their hands when they meet. More correctly, the system is based on an enhancement of the SPINE framework called Collaborative-SPINE (C-SPINE, see Chapter 7) and integrates handshake gesture detection with continuous beat-to-beat HR computation. This integrated information is useful for detecting emotional states of meeting people, when the meeting starts with a handshake.

The system architecture, depicted in Figure 11.7, is composed of two layers, located on the coordinator and on the wearable sensor devices. At the sensor level, the main components are:

- The Heart Rate Sensor (HRSensor) component, running on a Shimmer2R node, is equipped with an ECG sensor-board to extract the HR. The HR estimation uses a 5-point moving average filter to smoothen the HR curve.
- The Hand Shaking Sensor (HSSensor) component, running on a Shimmer2R node placed on the right wrist of the monitored subject, acquires accelerometer data for handshake recognition. The HSSensor (i) samples at 100 Hz the 3-axial accelerometer included in the Shimmer2R, (ii) buffers the acquired data, (iii) performs on this data specific feature extraction (amplitude, standard deviation, zero crossing, average, total energy, and RMS), (iv) runs a decision tree-based classifier for the detection of potential handshake gestures, and (v) finally transmits the computed feature set when a potential handshake gesture is recognized. In particular, the features are calculated over a window of 32 samples with 50% overlap. Such parameters have been empirically estimated to trade-off fast detection and good classification accuracy.

At the coordinator level, E-Shake is developed for the Android OS and integrates two application components: (i) a handshake detection component that uses C-SPINE to recognize the handshake gesture and (ii) a heart-rate component providing beat-to-beat HR data. Specifically, the coordinator aligns the HR data with handshake classifications obtained from a joint classifier and keep tracks of the HR data (which will be input for the emotion detection

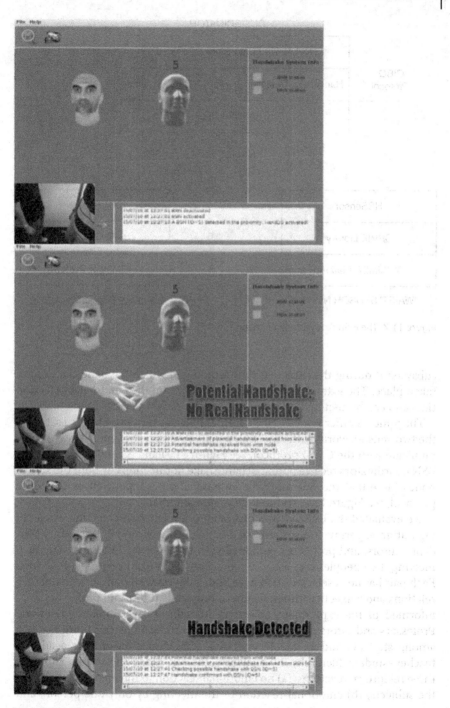

Figure 11.6 The E-Shake application.

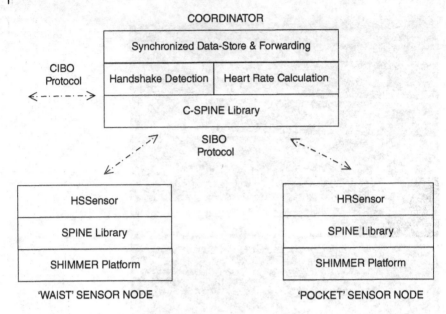

Figure 11.7 The E-Shake system architecture.

subsystem) during the extended time window in which the handshakes have taken place. The extended time window depends on the handshake detection time and can be centered on it or configured for asymmetric window sizing.

The joint classifier is a J48 decision tree that uses the entire feature set from the two sensors worn by the meeting people (the two BSN coordinators communicate with the CIBO protocol, see Figure 11.7) and is activated only if the BSN coordinators receive potential handshake notifications by both the sensor nodes (note that the "intra-BSN" communication is supported by the SIBO protocol, see Figure 11.7) within a short time interval.

We evaluated the E-Shake in terms of emotion reaction detection by carrying out an experimental scenario in a controlled environment, in which students, tutors, and professors equipped with the system could meet. For each meeting, two people were asked to enter the room from two separate doors. Each pair has been selected so to have both subjects with mutual acquaintance relations and subjects without. Furthermore, while tutors and professors were informed of the experiment objective, students were completely unaware. Professors and tutors played an important role to facilitate social interactions among students and as enablers of student reactions due to the academic teacher–student hierarchy. By analyzing the HR chart, the system captured three reciprocal reactions: (a) no emotional reaction to the meeting by none of the subjects, (b) emotional reaction to the meeting by only one person, and

(c) emotional reaction to the meeting by both meeting people. Cases similar to (a) are the most frequent, while cases similar to (b) are mostly associated with meetings between a student and a tutor or professor, and occasionally recorder for meetings among students. Occurrences of cases similar to (c) were actually very rare throughout the whole experiment.

11.7 Physical Rehabilitation

It is quite common to require repetitive physical exercises, for instance, to recover from a muscle strain, a limb fracture, or a surgery. Having real-time feedback about exercise performance quality would allow patients that are following a rehabilitation therapy to independently exercise correctly without the need of a continuous professional assistance.

11.7.1 Related Work

Although the literature on physical rehabilitation assistance supported by wearable sensors is still limited, a few interesting research studies have been published so far.

An early research [45] focuses on the therapist perspective aiming at determining the physical activity stress and the energy expenditure of therapists while practicing using a portable accelerometer sensor placed on their waist belts.

In Ref. [46], the authors propose the use of wearable accelerometer sensors for objectively assessing the motion capabilities and activity levels of patients affected by multiple sclerosis, so as not to rely uniquely on self-reports and questionnaires.

However, the specific problem of supporting patients during rehabilitation exercises with the aid of wearable sensing devices and real-time visual feedbacks is being investigated only in more recent times. In Ref. [47], the authors describe a rehabilitation support system based on a smartphone and a bracelet to capture patient's rehabilitation exercises. Dynamic Time Warping is used to train and recognize movements. The system is fully customizable so it allows the therapist to choose the position of the device and other parameters in order to adapt to different exercises. The proposed system, however, since relying on a single sensing device, suffers from the problem that a number of exercises cannot be monitored, and relevant parameters, such as elbow and knee flexion angles, cannot be measured.

RIABLO [48] is a game system realized to specifically support physical orthopedic rehabilitation. The authors suggest the use of game elements to motivate and engage the patient, while providing feedback on the correctness of the performed exercises. The system is based on five wearable devices equipped with a

3-axis accelerometer and a gyroscope, placed on the body with elastic straps, and a pressure sensor tile connected via Bluetooth with the game station.

Another interesting project [49] uses two Shimmer motes [9] attached to the patient's arm or leg and a commercial Android tablet where a graphical application provides with a visual real-time feedback on the performed exercises as well as an assessment on the practice quality with respect to a reference movement previously recorded.

In addition to purely academic research studies, there exist some pre-commercial solutions [50, 51] with similar functionalities to what described above.

For further literature study, readers can refer to interesting surveys [52, 53] recently published.

11.7.2   SPINE Motor Rehabilitation Assistant

In this section, we present a physical rehabilitation digital assistant (see Figure 11.8) that is implemented atop SPINE and uses two wearable nodes equipped with accelerometer sensors to monitor arm and leg movements. A personal mobile application runs on the patient's smartphone or tablet (Android-based) and gives real-time feedback on the performed exercise; furthermore, it interacts with a dedicated cloud computing backend to transfer collected data for long-term, off-line analysis and for retrieving comments

Figure 11.8 Two screenshots of the rehabilitation digital assistant.

and updates from the therapist about the rehabilitation process (e.g. to download the exercises schedule for the week).

The application consists of monitoring leg and arm bending movements in real time and comparing them with the ones recorded during a setup phase. The application scenario, hence, consists of two steps, namely setup and exercise phases. During the setup phase, the user wears two sensors on either leg or arm that needs to be exercised and performs the correct exercise under the guidance of rehabilitation professional. Meanwhile, the system records the data and stores it as reference exercise. Then, during the exercise phase, the user repeats the bending movement and is provided with a real-time feedback about how the movement is done with respect to the stored reference exercise.

Tele-rehabilitation is a crucial aspect, concerning the possibility to monitor the patient remotely. This possibility addresses, in particular, certain types of rehabilitation. This aspect is crucial since the idea of tying the opportunity to follow and monitor the patient at all post-admission stages through remote monitoring allows the substantial reduction of the costs associated with the process. As an example, we can consider the orthopedic rehabilitation of patients following violent trauma that are released from the hospital and proceed towards rehabilitation phases with low necessity of (constant) clinical doctor monitoring. In case of orthopedic trauma, the patient that can safely perform rehabilitation and can be remotely monitored will meet benefits both in physical stress and economic terms.

In addition, the system allows for the secure and authenticated collection via the Internet of data related to the management and control of rehabilitation by means of a dedicated cloud computing infrastructure. This cloud computing backend system is mainly intended to provide support to physicians. The doctors, through a web application are enabled to:

- Management of patients and their program exercises.
- Displaying data on the exercises performed by patients. The doctor must be able to analyze the exercises performed by his patients to be able to perceive the improvements that the prescribed therapy is expected to achieve. This greatly facilitates his work: thanks to the accuracy of the data, the therapist is "virtually" able to follow all his patients, just as if they were present in the rehabilitation center. If necessary, he may request a new appointment when he considers appropriate to change or update the prescribed therapy, and the patient is notified through the application running on his personal mobile device.
- Viewing statistics on the progress of patients. Doctors need quick and easy-to-read information related to the progress of his patient throughout the therapy period. The doctor is supported in this sense by synthetic statistics such as maximum and minimum extension and flexion angles (of the elbow or the knee), torsion angles (of the leg or the arm), range of motion, and minutes of training per day.

11.8 Summary

The main goal of SPINE is to provide BSN developers with support for rapid prototyping of signal-processing applications. In SPINE, sensors and common processing blocks, such as math aggregators and threshold-based alarms, can be configured independently and connected together arbitrarily at runtime based on external controls. One of the key advantages of SPINE is the ability to satisfy diverse application needs at runtime, avoiding, in most situations, the costly redeployment of the code running on the remote sensing devices.

Such an approach also allows heterogeneous applications to be built atop the same basic software components, enhancing code reusability and, more importantly, removes the need for redeploying the node-side code based on a particular application.

This property is very desirable especially in real-world scenarios. For instance, a doctor could use SPINE nodes that are equipped with accelerometers and a suitable coordinator device (e.g. a smartphone) to monitor weekly energy expenditure of a patient. The same nodes could be used later with another patient, for instance, in a rehabilitation scenario, as long as the proper application software is available on the doctor's coordinator device. In this chapter, the SPINE framework has been showed to support heterogeneous health-care applications without redeployment of the code running on the nodes. The flexibility of SPINE has been demonstrated by describing five different case studies (physical activity detection, step counting, emotional stress detection, handshake detection, and physical rehabilitation), which all exploit the same sensor node hardware and software. Obviously, in the general case, to support different applications, the wearable sensing node(s) must be equipped with all the required physical sensors.

References

1 Bao, L. and Intille, S.S. (2004). Activity recognition from user-annotated acceleration data. *Pervasive 2004*, LNCS 3001, Linz/Vienna, Austria (18–23 April), pp. 1–17.

2 Maurer, U., Smailagic, A., Siewiorek, D., and Deisher, M. (2006). Activity recognition and monitoring using multiple sensors on different body positions. *Proceedings of the International Workshop on Wearable and Implantable Body Sensor Networks*, Cambridge, MA (3–5 April 2006), pp. 99–102.

3 Ravi, N., Preetham Mysore, N.D., and Littman, M.L. (2005). Activity recognition from accelerometer data. *Proceedings of the 17th Conference on Innovative Applications of Artificial Intelligence*, Pittsburgh, PA (9–13 July 2005), pp. 1541–1546.

4 Zappi, P., Lombriser, C., Stiefmeier, T. et al. (2008). Activity recognition from on-body sensors: accuracy-power trade-off by dynamic sensor selection.

Proceedings of the European Conference on Wireless Sensor Networks, Bologna, Italy (30 January–1 February 2008), pp. 17–33.

5 Anguita, D., Ghio, A., Oneto, L. et al. (2012). Human activity recognition on smartphones using a multiclass hardware-friendly support vector machine. *Proceedings of the 4th international conference on Ambient Assisted Living and Home Care*, Vitoria-Gasteiz, Spain, 3–5 December 2012, pp. 216–223.

6 Shoaib, M. (2013). Human activity recognition using heterogeneous sensors. *Proceedings of ACM International Joint Conference on Pervasive and Ubiquitous Computing*, Zurich, Switzerland (8–12 September 2013).

7 Chen, L., Hoey, J., and Nugent, C.D. (2012). Sensor-based activity recognition. *IEEE Transactions on Systems, Man, and Cybernetics, Part C: Applications and Review* 42 (6): 790–808.

8 Lara, O.D. and Labrador, M.A. (2013). A survey on human activity recognition using wearable sensors. *IEEE Communications Surveys & Tutorials* 15 (3): 1192–1209.

9 Shimmer Website. www.shimmersensing.com (accessed 15 June 2017).

10 Cover, T. and Hart, P. (1997). Nearest neighbor pattern classification. *IEEE Transactions on Information Theory* 13: 21–27.

11 Pudil, P., Novovicova, J., and Kittler, J. (1994). Floating search methods in feature selection. *Pattern Recognition Letters* 15 (11): 1119–1125.

12 Carter, B.C., Vershinin, M., and Gross, S.P. (2008). A comparison of step-detection methods: how well can you do? *Biophysical Journal* 94 (1): 306–319.

13 Oliver, M., Badland, H.M., Shepherd, J., and Schofield, G.M. (2011). Counting steps in research: a comparison of accelerometry and pedometry. *Open Journal of Preventive Medicine* 1: 1–7.

14 Libby, R. (2008). A simple method for reliable footstep detection on embedded sensor platforms. https://www.scribd.com/document/136324023/Libby-Peak-Detection (accessed 6 December 2017).

15 Ahola, T.M. (2010). Pedometer for running activity using accelerometer sensors on the wrist. *Medical Equipment Insights* 2010 (3): 1–8.

16 Wu, L.-M., Sheu, J.-S., Jheng, W.-C., and Hsiao, Y.-T. (2013). Pedometer development utilizing an accelerometer sensor. *World Academy of Science, Engineering and Technology* 79: 35–40.

17 McEwen, B.S. (1998). Protective and damaging effects of stress mediators. *The New England Journal of Medicine* 338 (3): 171–179.

18 Segerstrom, S.-C. and Miller, G.-E. (2004). Psychological stress and the human immune system: a meta-analytic study of 30 years of inquiry. *Psychological Bulletin* 130 (4): 601–630.

19 Lee, H.B., Kim, J.S., Kim, Y.S. et al. (2007). The relationship between HRV parameters and stressful driving situation in the real road. *Proceedings of the 6th International Special Topic Conference on Information Technology Applications in Biomedicine*, Tokyo, Japan (8–11 November 2007), pp. 198–200.

20 Sun, F.-T., Kuo, C., Cheng, H.-T. et al. (2012). Activity-aware mental stress detection using physiological sensors. *Lecture Notes of the Institute for Computer Sciences, Social Informatics and Telecommunications Engineering* 76, 211–230.

21 de Santos Sierra, A., Sánchez Ávila, C., Guerra Casanova, J., and del Pozo, G.B. (2011). Real-time stress detection by means of physiological signals. In: *Recent Application in Biometrics* (ed. J. Yang and N. Poh), 23–44. London: Intech.

22 Stress Eraser Website. www.stresseraser.com (accessed 15 June 2017).

23 Health Reviser Stress Monitor. http://www.healthreviser.com/content/stress-sweeper (accessed 6 December 2017).

24 emWave Personal Stress Reliever. https://store.heartmath.com/emwave2 (accessed 15 June 2017).

25 Andreoli, A., Gravina, R., Giannantonio, R. et al. (2010). SPINE-HRV: a BSN-based toolkit for heart rate variability analysis in the time-domain. *Wearable and Autonomous Biomedical Devices and Systems: New Issues and Characterization – Lecture Notes on Electrical Engineering* 75: 369–389.

26 Polar Website. www.polar.com (accessed 15 June 2017).

27 Yang, H.-K., Lee, J.-W., Lee, K.-H. et al. (2008). Application for the wearable heart activity monitoring system: analysis of the autonomic function of HRV. *Proceedings of the 30th Annual International Conference on Engineering in Medicine and Biology Society (EMBS 2008)*, Vancouver, Canada (20–25 August 2008), pp. 1258–1261. IEEE Press.

28 Bauer, R. (1998). Physiologic measures of emotion. *Journal of Clinical Neurophysiology* 15 (5): 388–396.

29 Lopez, R., Poy, R., Pastor, M. et al. (2009). Cardiac defense response as a predictor of fear learning. *International Journal of Psychophysiology* 74 (3): 229–235.

30 Zimbardo, P.G., Weber, A.L., and Johnson, R.L. (1999). *Psychology*, 3e. Boston, MA: Addison Wesley Longman, Ed.

31 Vila, J., Fernandez, M.C., Pegalajar, J. et al. (2003). A new look at cardiac defense: attention or emotion? *The Spanish Journal of Psychology* 6 (1): 60–78.

32 Sebe, N., Cohen, I., Gevers, T., and Huang, T.S. (2004). Multimodal approaches for emotion recognition: a survey. *Internet Imaging VI* 5670: 56–67.

33 Cowie, R., Douglas-Cowie, E., Tsapatsoulis, N. et al. (2001). Emotion recognition in human-computer interaction. *IEEE Signal Processing Magazine* 18 (1): 32–80.

34 Martyn Jones, C. and Troen, T. (2007). Biometric valence and arousal recognition. *Proceedings of the 2007 Conference of the Computer-Human Interaction Special Interest Group (CHISIG) of Australia on Computer-Human Interaction*, Adelaide, Australia (28–30 November 2007).

35 Grundlehner, B., Brown, L., Penders, J., and Gyselinckx, G. (2009). The design and analysis of a real-time, continuous arousal monitor. *Sixth International*

Workshop on Wearable and Implantable Body Sensor Networks, Berkeley, CA (3–5 June 2009), pp. 156–161.

36 Valenza, G., Lanatà, A, Scilingo, E.P., and De Rossi, D. (2010). Towards a smart glove: arousal recognition based on textile electrodermal response. *2010 Annual International Conference of the IEEE Engineering in Medicine and Biology Society*, Buenos Aires, Argentina (31 August 2010–4 September 2010), pp. 3598–3601.

37 Covello, R., Fortino, G., Gravina, R. et al. (2013). Novel method and real-time system for detecting the Cardiac Defense Response based on the ECG. *Proceedings of the IEEE International Symposium on Medical Measurement and Applications (MeMeA2013)*, Trento, Italy (16–20 September 2013).

38 Gravina, R. and Fortino, G. (2016). Automatic methods for the detection of accelerative cardiac defense response. *IEEE Transactions on Affective Computing* 7 (3): 286–298.

39 Chen, H. and Chen, S. (2003). A moving average based filtering system with its application to real-time QRS detection. *Proceedings of Computers in Cardiology, ser. CinC 2003*, Thessaloniki Chalkidiki, Greece (21–24 September 2003), pp. 585–588.

40 Kiremire, B. and Marwala, T. (2008). Nonstationarity detection: the use of the cross correlation integral in ECG, and EEG profile analysis. *Proceedings of the Congress on Image and Signal Processing, ser. CISP'08*, Singapore (27–30 May 2008), pp. 373–378.

41 Rserve Website. http://www.rforge.net/Rserve (accessed 15 June 2017).

42 Kanis, M., Winters, N., Agamanolis, S. et al. (2005). Toward wearable social networking with iband. *Proceedings of Computer-Human Interaction (CHI)* – Extended abstracts on Human factors incomputing systems, Portland, OR (2–7 April 2005), pp. 1521–1524. ACM.

43 Holmquist, L.E., Mattern, F., Schiele, B. et al. (2001). Smart-its friends: a technique for users to easily establish connections between smart artefacts. *Proceedings of the 3rd International Conference on Ubiquitous Computing (UbiComp)*, Atlanta, GA (30 September–2 October 2001), pp. 116–122. Springer-Verlag.

44 Augimeri, A., Fortino, G., Galzarano, S., and Gravina, R. (2011). Collaborative body sensor networks. *Proceedings of the IEEE International Conference on Systems, Man and Cybernetics (SMC2011)*, Anchorage, AL (9–12 October 2011).

45 Balogun, J.A., Farina, N.T., Fay, E. et al. (1986). Energy cost determination using a portable accelerometer. *Physical Therapy* 66: 1102–1107.

46 Hale, L., Williams, K., Ashton, C. et al. (2007). Reliability of RT3 accelerometer for measuring mobility in people with multiple sclerosis: pilot study. *Journal of Rehabilitation Research & Development* 44 (4): 619–628.

47 Raso, I., Hervás, R., and Bravo, J. (2010). m-Physio: personalized accelerometer-based physical rehabilitation platform. *Proceedings of the 4th*

International Conference on Mobile Ubiquitous Computing, Systems, Services and Technologies, Florence, Italy (25–30 October 2010).

48 Costa, C., Tacconi, D., Tomasi, R. et al. (2013). RIABLO: a game system for supporting orthopedic rehabilitation. *CHItaly 2013, the Biannual Conference of the Italian SIGCHI Chapter*, Trento, Italy (16–20 September 2013).

49 Nerino, R., Contin, L., Gonçalves da Silva Pinto, W.J. et al. (2013). A BSN based service for post-surgical knee rehabilitation at home. *Proceedings of the 8th International Conference on Body Area Networks*, Boston, MA (30 September–2 October 2013).

50 Rehabitic Whitepaper. http://www.imim.es/media/upload/arxius/oferta% 20tecnologica/REHABITICwebIMIM_EN.pdf (accessed 15 June 2017).

51 PamSys Website. www.biosensics.com (accessed 15 June 2017).

52 Patel, S., Park, H., Bonato, P. et al. (2012). A review of wearable sensors and systems with application in rehabilitation. *Journal of NeuroEngineering and Rehabilitation* 9: 21.

53 Hadjidj, A., Souil, M., Bouabdallaha, A. et al. (2013). Wireless sensor networks for rehabilitation applications: challenges and opportunities. *Journal of Network and Computer Applications* 36: 1–5.

12

SPINE at Work

12.1   Introduction

This chapter provides a quick yet effective reference for BSN programmers interested in developing their applications using the SPINE framework. While a comprehensive developer guide can be freely downloaded from the website, in this chapter we will give the necessary information for setting up the SPINE environment so to start programming as well as insights on how the framework itself can be customized and extended.

12.2   SPINE 1.x

SPINE (Signal Processing In-Node Environment) (see Chapter 3) is a framework for the distributed implementation of signal-processing algorithms in wireless sensor networks.

It provides a set of on-node services that can be tuned and activated by the user depending on application needs.

SPINE is released as an Open Source project under LGPL 1.2 license and is available online at http://spine.dimes.unical.it/.

The SPINE framework has two main components:

1) *Sensor node side.* It is developed in the TinyOS2.x environment and provides on-node services such as sensor data sampling and storage, data processing, and much more.
2) *Server side.* It is developed in Java SE and acts as a coordinator of the sensor networks. Therefore, it manages the network, setups and activates on-node services depending on the application requirements, and much more.

The framework has been redesigned and the newest release (1.3) provides many more levels of expansibility than the previous releases.

Wearable Computing: From Modeling to Implementation of Wearable Systems Based on Body Sensor Networks, First Edition. Giancarlo Fortino, Raffaele Gravina, and Stefano Galzarano.
© 2018 John Wiley & Sons, Inc. Published 2018 by John Wiley & Son, Inc.

The core framework is now organized into three main parts that take care of different aspects, namely the communication, the sensing, and the processing parts.

The source code of the node side is organized as follows:

```
Spine_nodes
|__apps
| |__SPINEApp
|__support
| |__make
|__tos
| |__interfaces
| | |__communication
| | |__processing
| | |__sensing
| | |__utils
| |__platforms
| |__sensorboards
| |__system
| | |__communication
| | |__processing
| | |__sensing
| | |__utils
| |__types
```

The Server side is organized as follows:

```
Spine_serverApp
|
|__src
| |
| |__jade.util
| |__spine
| | |
| | |__communication.emu
| | |__communication.tinyos
| | |
| | |__datamodel
| | |
| | |__datamodel.functions
| | |__datamodel.serviceMessages
| | |__exceptions
| | |__payload.codec.emu
```

```
| | |__payload.codec.tinyos
| |
| |__test
|
|__lib
|
|__jar
|
|__doc
|
|__resources/defaults.properties
|
|__build.xml
|
|__build.prope rties
```

This structure reflects the need of having the framework logic not depending on the kind of network it is communicating with. In other words, the core implementation of SPINE does not use any TinyOS-specific APIs and can be run independently on the underlying protocol stack (e.g. ZigBee networks). Platform-independent code may be found in:

- spine package, which contains SPINE core logic.
- spine.datamodel package, which contains data entities used by the framework.
- spine.datamodel.functions subpackage, which defines the structure of the function.
- spine.datamodel.serviceMessages subpackage, which defines various types of service messages.
- spine.exceptions subpackage, which contains exception classes that might be thrown by SPINE.

SPINE1.3 server side provides an implementation for the TinyOS2.x network and for the "virtual sensor node" network; therefore it provides the support for TinyOS low-level communication:

- spine.communication.tinyos contains TinyOS-specific logic and low-level IEEE 802.15.4-based communication procedures (called tinyos.jar APIs).
- spine.payload.codec.tinyos subpackage contains the low-level message codecs for the TinyOS platform.
- spine.communication.bt contains low-level Bluetooth-based communication procedures (using the open-source BlueCove library on desktop computers and the native Bluetooth API on Android).
- spine.payload.codec.bt subpackage contains the low-level message codecs for Bluetooth serial transmission.

For "SPINE Node Emulator" (each "Node Emulator" instance is a "virtual sensor node"; see Data Collector and SPINE Node Emulator) low-level communication:

- `spine.communication.emu` contains logic and low-level communication procedures for virtual sensor node.
- `spine.payload.codec.emu` subpackage contains the low-level message codecs for the virtual sensor node message.

SPINE1.3 release provides also the SPINE.jar that can be imported in any project that uses SPINE APIs and the full javadoc documentation.

12.2.1 How to Install SPINE 1.x

Installing SPINE onto the target platforms is straightforward. The process consists of the following steps:

1) Download SPINE 1.3 from the SPINE website (http://spine.dimes.unical.it/). The unzipped spine folder contains:
   a) Spine_nodes folder with TinyOS2.x code to be run on the motes.
   b) Spine_serverApp folder with Java code to be run on a computer.
   c) COPYING and License text files containing info about the licensing.
   d) The SPINE manual.
2) Spine_nodes contains code to be compiled in TinyOS2.x and then flashed on sensor nodes. Spine_nodes 1.3 has been developed and tested with TinyOS version 2.1.0. Older TinyOS2.x versions have also been tested, and Makefile can be configured to support an older version, but the SPINE Team strongly suggests to use TinyOS2.1.0 release.
   a) Copy Spine_nodes folder into your tinyos-2.x-contrib folder.
   b) From the app/SPINEApp folder compile and install the SPINE1.3 framework on your platform. Platforms currently supported by SPINE1.3 are:
      i) Telosb motes with spine sensor board
         `SENSORBOARD=spine make telosb`
      ii) Telosb motes with biosensor sensor board
         `SENSORBOARD=biosensor make telosb`
      iii) Telosb motes with the moteiv sensor kit
         `SENSORBOARD=moteiv make telosb`
      iv) Micaz motes with mts300 board
         `SENSORBOARD=mts300 make micaz`
      v) Shimmer motes
         `SENSORBOARD=shimmer make shimmer`
      vi) Shimmer2 motes
         `SENSORBOARD=shimmer2 make shimmer2`
      vii) Shimmer2r motes
         `SENSORBOARD=shimmer2r make shimmer2r`

Note that for each supported platform, a default SENSORBOARD has been defined. Therefore, unless differently specified (e.g. by defining the SENSORBOARD parameter in the make command):

- telosb defaults to "spine" sensorboard.
- tmote defaults to "moteiv" sensorboard.
- micaz defaults to "mts300" sensorboard.
- shimmer defaults to "shimmer" sensorboard.

To change these default values, the corresponding details can be found in the tos/types/spine.extra file.

3) Spine_serverApp contains the Java code for running the server side (e.g. coordinator) of a SPINE network.
   a) src contains SPINE1.3 source code organized into:
      - spine
      - jade
      - test
   b) defaults.properties contains the framework properties.
   c) lib: contains a jar file that SPINE must include.
   d) docs: contains SPINE1.3 javadoc documentation.
   e) jar: contains the framework jar file.
   f) build.properties and build.xml files for ant.

It is possible to compile and run the SPINE framework and its test application either using textual ant commands or creating a java project using an IDE (such as Eclipse or NetBeans). An external jar (tinyos.jar) has to be manually added to the project. Due to different copyright regulations, this jar is not part of the SPINE distribution and can be found in the tinyos2.x\ support\sdk\java folder. tinyos.jar should be placed in the spine_serverApp/ ext-lib folder.

12.2.2 How to Use SPINE

The SPINE framework provides, on the Server side, simple Java APIs to develop applications on the coordinator. Therefore, the main strength of the SPINE framework is to allow users to be ready to develop applications in sensor networks without bothering with node-side programming.

Developers can easily form, manage, and collect data from the sensors in the network writing a simple Java program: no more firmware programming is needed!

On the Java side, the user can develop its own application that will have to implement the SPINEListener interface and can use any of the API provided by the SPINEManager.

Since the application on the server side must implement the SPINEListener interface, it has to implement the following methods:

void **received**(Data data)
This method is invoked by the SPINEManager to its registered listeners when it receives new data from the specified node. The Node object that generated this data is embodied into the data object.

void **discoveryCompleted**(java.util.Vector activeNodes)
This method is invoked by the SPINEManager to its registered listeners when the discovery procedure timer fires.

void **newNodeDiscovered**(Node newNode)
This method is invoked by the SPINEManager to its registered listeners when it receives a ServiceAdvertisement message from a BSN node.

void **received**(ServiceMessage msg)
This method is invoked by the SPINEManager to its registered listeners when a ServiceMessage is received from a particular node. The Node object that generated this service message is embodied into the msg object.

Then, the application can use the following API exposed by the SPINEManager:

void **activate**(Node node, SpineFunctionReq functionReq)
Activates a function (or even only function subroutines) on the given sensor.

void **addListener**(SPINEListener listener)
Registers a SPINEListener to the manager instance.

void **deactivate**(Node node, SpineFunctionReq functionReq)
Deactivates a function (or even only function subroutines) on the given sensor.

void **discoveryWsn**()
Commands the SPINEManager to discover the surrounding WSN nodes.

java.util. **getActiveNodes**()
Vector
Returns the list of the discovered nodes as a Vector of spine.datamodel.Node objects.

spine.data **getBaseStation**()
model.Node
Returns the Node object representing the BaseStation.

Jade.util. Logger	**static getLogger**() Returns the static Logger of the SPINE framework. The Logger can be used to set the logging level and to add custom log handlers (e.g. to log into a file).
spine.data model.Node	**getNodeByLogicalID**(spine.datamodel. Address id) Returns the node with the given logical address.
spine.data model.Node	**getNodeByPhysicalID** (spine.datamodel. Address id) Returns the node with the given physical address.
void	**getOneShotData**(Node node, byte sensorCode) Commands the given node to do an "immediate one- shot" sampling on the given sensor.
boolean	**isStarted**() Returns true if the manager has been asked to start the processing in the BSN.
void	**removeListener**(SPINEListener listener) Removes a SPINEListener from the manager instance.
void	**reset**() Commands a software reset of the whole WSN.
void	**setup**(Node node, SpineSetupFunction setupFunction) Setups a specific function of the given node.
void	**setup**(Node node, SpineSetupSensor setupSensor) Setups a specific sensor of the given node.
void	**start**(boolean radioAlwaysOn, boolean enableTDMA) Starts the BSN sensing and computing the previously requested functions.

It is worth noting that the SPINEManager instance can be retrieved only via the SPINEFactory:

SPINEManager	**createSPINEManager**(String appPropertiesFile) Initializes the SPINEManager. The SPINEManager instance is connected to the base station and platform obtained transparently from the app.properties file.

Examples about which function can be set, which data can be received, and other details can be found in the SPINETest application, that is included in the latest release of the SPINE source code. More examples about how to use the Java side are given all through this document.

For further details about the Java side, the interested reader can refer to the Javadoc documentation that can be found in the release.

12.2.3 How to Run a Simple Desktop Application Using SPINE1.3

The SPINE1.3 release comes with a simple test application that can be easily run to experiment the framework basic functionalities. Take the following steps:

1) Compile and flash, on the available sensor node, the SPINE1.3 node-side framework.
2) Compile and flash a TinyOS2.x BaseStation into another sensor node. It is important to check that sensor nodes and the base station are both working on the same radio channel, have been compiled with the same max message payload length, and the same TinyOS version has been used for flashing all the nodes.
3) Plug the BaseStation to a free USB port of the computer and type "motelist" from your shell: this will tell you your port number.
4) Create an application properties file (e.g. under MyApp/resources/app. properties) and set the MOTECOM and the PLATFORM parameter according to one of the following options depending if you are using the serial for-warder on a Linux or Windows machine (e.g. a.) or directly communicating with the serial port on your PC using a Windows machine (e.g. b.), or if you intend to emulate a sensor node network (e.g. c.).
    a) MOTECOM=sf@127.0.0.1:9002
       PLATFORM=sf
    b) MOTECOM=serial@COM41:telosb
       PLATFORM=tinyos
    c) MOTECOM=4444
       PLATFORM=emu

Option b may be used also on a Linux machine, but it is necessary to build *libgetenv* and *libtoscomm* library before being able to install and run any SPINE application. Also, MOTECOM value would look like "serial@/dev/ttyS0:telosb."

```
cd $TOSROOT/support/sdk/java && make
     sudo tos-install-jni
```

If needed, other application-dependent properties can be stored in this property file without any side effect to the SPINE framework.

5) edit Spine_serverApp/test/SPINETest.java and optionally go through the code to customize the test application. The code documentation helps to understand what functionalities SPINE exposes to the java developer.

As mentioned before, SPINETest.java implements the SPINEListener interface (to get notified of SPINE-related events) and uses the SPINEFactory to retrieve the SPINEManager, which, in turn, has the APIs for managing and communicating with the nodes in the network.

The SPINETest provided within the SPINE 1.3 release performs the following actions:

a) a discovery message is broadcasted to check how the PAN is composed:
   ```
   manager.discoveryWsn();
   ```
b) when the discovery is completed, all the received info about nodes present in the PAN is displayed.

   ```
   curr = (Node)activeNodes.elementAt(j);
   // we print for each node its details (nodeID,
   sensors, and functions provided)
   System.out.println(curr);
   ```

   The information displayed at this point is:

   i)   node id
   ii) supported sensors
   iii) supported functionalities
c) if a node with an accelerometer is found:
   i)   the accelerometer is set with sampling time SAMPLING_TIME=50 ms.

   ```
   SpineSetupSensor sss = new SpineSetupSensor();
   sss.setSensor(sensor);
   sss.setTimeScale(SPINESensorConstants.MILLISEC)
   sss.setSamplingTime(SAMPLING_TIME);
   manager.setup(curr, sss);
   ```

   ii) the feature engine function is set on that node to work on data coming from the accelerometer sensor with window WINDOW_SIZE=40 and shift SHIFT_SIZE=20.

   ```
   FeatureSpineSetupFunction ssf = new FeatureSpine
   SetupFunction();
   ssf.setSensor(sensor);
   ssf.setWindowSize(WINDOW_SIZE);
   ```

```
ssf.setShiftSize(SHIFT_SIZE);
manager.setup(curr, ssf);
```

iii) few features are activated on that node on the accelerometer data (MODE, MEDIAN, MAX, and MIN on all the accelerometer's channels).

```
FeatureSpineFunctionReq sfr = new
FeatureSpineFunctionReq();
sfr.setSensor(sensor);
sfr.add(new Feature(SPINEFunctionConstants.MODE,
((Sensor)curr.getSensorsList().elementAt(i))
    .getChannelBitmask()));
sfr.add(new Feature(SPINEFunctionConstants.MEDIAN,
((Sensor)curr.getSensorsList().elementAt(i))
    .getChannelBitmask()));
sfr.add(new Feature(SPINEFunctionConstants.MAX,
((Sensor)curr.getSensorsList().elementAt(i))
.getChannelBitmask()));
sfr.add(new Feature(SPINEFunctionConstants.MIN,
((Sensor) curr.getSensorsList().elementAt(i))
.getChannelBitmask()));
manager.activate(curr, sfr);
```

iv) more features are activated (MEAN, AMPLITUDE).

```
FeatureSpineFunctionReq sfr = new
FeatureSpineFunctionReq();
sfr.setSensor(sensor);
sfr.add(new Feature(SPINEFunctionConstants.MEAN,
((Sensor) curr.getSensorsList().elementAt(i))
.getChannelBitmask()));
sfr.add(new Feature(SPINEFunctionConstants.
AMPLITUDE,
((Sensor) curr.getSensorsList().elementAt(i))
.getChannelBitmask()));
manager.activate(curr, sfr);
```

v) the alarm engine function is set on the node to work on data coming from the accelerometer sensor with window WINDOW_SIZE=40 and shift SHIFT_SIZE=20. Note that Feature and Alarm engines can be set with different settings, since they are two separate components. However, in this test application, they have been set with the same value to better check the results.

```
AlarmSpineSetupFunction ssf2 = new
AlarmSpineSetupFunction();
```

```
ssf2.setSensor(sensor);
ssf2.setWindowSize(WINDOW_SIZE);
ssf2.setShiftSize(SHIFT_SIZE);
manager.setup(curr, ssf2);
```

vi) two alarms are set on the accelerometer sensor, so that an alarm message will be sent back when:

    1) the MAX value on CH1 is greater than upperThreshold value = 40.

```
AlarmSpineFunctionReq sfr2 = new
AlarmSpineFunctionReq();
sfr2.setDataType(SPINEFunctionConstants.MAX);
sfr2.setSensor(SPINESensorConstants.
ACC_SENSOR);
sfr2.setValueType((SPINESensorConstants.
CH1_ONLY));
sfr2.setLowerThreshold(lowerThreshold);
sfr2.setUpperThreshold(upperThreshold);
sfr2.setAlarmType(SPINEFunctionConstants.
ABOVE_THRESHOLD);
manager.activate(curr, sfr2);
```

    2) the AMPLITUDE on CH2 is lower than lowerThreshold value = 2000.

```
sfr2.setDataType(SPINEFunctionConstants.AMPLITUDE);
sfr2.setSensor(SPINESensorConstants.ACC_SENSOR);
sfr2.setValueType((SPINESensorConstants.
CH2_ONLY));
sfr2.setLowerThreshold(lowerThreshold);
sfr2.setUpperThreshold(upperThreshold);
sfr2.setAlarmType(SPINEFunctionConstants.
BELOW_THRESHOLD);
manager.activate(curr, sfr2);
```

d) if a node with internal CPU temperature sensor is found:

    i) the temperature sensor is set with sampling time OTHER_SAMPLING_TIME=100 ms.

    ii) the feature engine function is set on that node to work on data coming from the temperature sensor with window OTHER_WINDOW_SIZE=80 and shift OTHER_SHIFT_SIZE=40.

    iii) few features are activated on that node on the temperature data (MODE, MEDIAN, MAX, and MIN).

    iv) the alarm engine function is set on the node to work on data coming from the accelerometer sensor with window WINDOW_SIZE=40 and shift SHIFT_SIZE=20.

v) then one alarm is set on the internal CPU temperature sensor, so that an alarm message will be sent back when the MIN value on CH1 is greater than 1000 and lower than 3000.

e) once all the requests are set, the network starts.

```
manager.startWsn(true, true);
```

f) on reception of the activated data (received(Data data)), data payload is displayed.

```
System.out.println(data);
```

g) during application runtime, functions can be deactivated and activated. Here for instance:

i) After receiving five feature packets, the first activated feature on that sensor is deactivated.

```
if(counter == 5) {
   // it's possible to deactivate functions computation
   at runtime (even when the radio on the node works
   in low-power mode)
   FeatureSpineFunctionReq sfr = new
   FeatureSpineFunctionReq();
   sfr.setSensor(features[0].getSensorCode());
   sfr.remove(new Feature(features[0].
getFeatureCode(),
   SPINESensorConstants.ALL);
   manager.deactivate(data.getNode(), sfr));
}
```

ii) After receiving 10 feature packets a new feature (RANGE) is computed on the first channel of that sensor.

```
if(counter 3== 10) {
   // and we can activate new functions at runtime
   FeatureSpineFunctionReq sfr = new
   FeatureSpineFunctionReq();
   sfr.setSensor(features[0].getSensorCode());
   sfr.add(new Feature(SPINEFunctionConstants.
RANGE,
   SPINESensorConstants.CH1_ONLY);
   manager.activate(data.getNode(), sfr);
}
```

iii) After 20 alarm packets the, the alarm – previously set to fire when the MAX value on CH1 is above the threshold value – is disabled.

```
if (counter_alarm == 20) {
   AlarmSpineFunctionReq sfr2 = new
   AlarmSpineFunctionReq();
   sfr2.setSensor(SPINESensorConstants.ACC_SENSOR);
   sfr2.setAlarmType(SPINEFunctionConstants.
   ABOVE_THRESHOLD);
   sfr2.setDataType(SPINEFunctionConstants.MAX);
   sfr2.setValueType((SPINESensorConstants.
   CH1_ONLY));
   manager.deactivate(data.getNode(), sfr2);
}
```

12.2.4   SPINE Logging Capabilities

The SPINE framework uses a Logger to print info or warning messages, to notify of exceptions, and so on. This enables a convenient way to filter undesired messages, to forward logs to output files, and much more.

From a SPINE user point of view, it can be useful to use the SPINE-Manager static method getLogger(), e.g. to modify the default logging level (INFO):

```
SPINEManager.getLogger().setLevel(Level.WARNING);
```

From a SPINE developer point of view, it is worth to report the correct way to print using the logger:

```
if (SPINEManager.getLogger()
        .isLoggable(Logger.[SEVERE|WARNING|INFO]))
SPINEManager.getLogger().log(Logger.
[SEVERE|WARNING|INFO], "msg");
```

Logging levels are hierarchical in terms of gravity. For instance, if the logging level has been set to WARNING, only the SEVERE and the WARNING messages will be logged, while INFO messages will not.

The interested reader can refer to the Jade Framework logging tutorial (http://jade.tilab.com/doc/tutorials/logging/JADELoggingService.html) and to the java.util.logging javadocs for further details.

12.3   SPINE2

SPINE2 (see Chapter 4) has not been conceived as a substitute of SPINE 1.x but is rather a parallel research effort aiming at (i) experimenting a different programming abstraction based on a task-oriented paradigm and (ii) designing a node-side software architecture for a quicker framework porting towards new sensor platforms.

Similar to SPINE, SPINE2 has two main components:

- A *(coordinator) server-side* management application (with GUI) and libraries providing functionalities and Java-API for (i) interfacing with the sensor network, (ii) defining and managing the task-based distributed application to be run on the sensor nodes, and (iii) feeding user-defined customized tools with the gathered data from the network for further data processing.
- A *sensor node* middleware providing sensing and distributed data processing functionalities by executing the tasks defined by the user. The middleware in turn is composed of two different sets of components: the core platform-independent modules, which can be in principle compiled for any C-like embedded platform with very slight changes in the source code, and the platform-dependent ones devoted to managing the physical resources and the lower level services provided by the specific platform. In this chapter, we specifically concentrate on the TinyOS port of SPINE2.

The platform-independent source code, i.e. the common node-side core framework, is organized as follows:

```
Spine2_common_c
|__actuating
|__communication
|__memory
|__sensing
|__task
|  |__task_list
|__timing
|__utils
|__SPINEManager.c
|__SPINEManager.h
```

In particular, the SPINEManager, placed in the root folder, is the central component of the core and is in charge of (i) system initialization and startup; (ii) orchestrating the other modules managing tasks, memory, sensors, actuators, and communication; and (iii) handling the SPINE2 application-level protocol. In the "task" folder are the modules for managing the task-graph representation and the scheduler for correctly instantiating and running the tasks allocated on the sensor node, whereas the "task_list" contains the library of tasks, i.e. the specifications of all the types of task supported by the framework. The "memory" folder has components in charge of dynamically allocating the task-based application as well as the buffers required by each single allocated task. The other folders contain modules for managing actuators and sensors, timers, communication, and providing other utility functions.

The TinyOS-specific node-side source code is organized as follows:

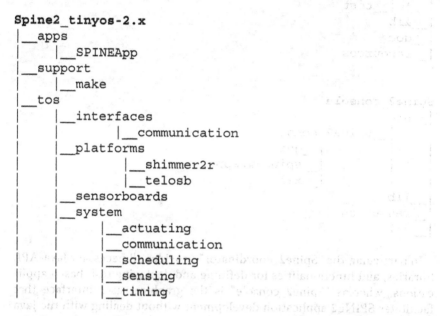

```
Spine2_tinyos-2.x
|__apps
|    |__SPINEApp
|__support
|    |__make
|__tos
|    |__interfaces
|    |         |__communication
|    |__platforms
|    |         |__shimmer2r
|    |         |__telosb
|    |__sensorboards
|    |__system
|    |         |__actuating
|    |         |__communication
|    |         |__scheduling
|    |         |__sensing
|    |         |__timing
```

Differently from the SPINE (version 1.x) source code, most of the source files merely contain "glue code" (i.e. adaptation components) binding the previously described SPINE2 functionalities with the TinyOS-specific sensor platforms code, which is used to access the lower level mechanisms and services (i.e. physical sensor/actuator, timer, and radio drivers). In particular, the "system" and "interfaces" folders contain TinyOS components related to the SPINE2 architecture, whereas "platforms" and "sensorboards" contain code binding with more specific drivers for sensor platforms and sensor boards.

The structure of the server-side management application (Java code) running on the coordinator of the sensor networks is organized as follows:

```
Spine2_coordinator
|__src
|    |__spine2
|    |       |__communication
|    |       |         |__tinyOS
|    |       |__message
|    |       |         |__message_list
|    |       |__support
|    |       |__task
|    |       |         |__task_list
|    |       |__utils
```

```
|        |          |__wsn
|        |__test
|__lib
|__doc
|__resources


Spine2_console
|__src
|        |__spine2.console
|        |          |__gui
|        |          |__spine2wrapper
|        |          |__xml
|__lib
|__resources
|__xml
```

In particular, the "Spine2_coordinator" provides all necessary Java-API, libraries, and functionalities for defining and deploying task-based applications, whereas "Spine2_console" is the graphical user interface that facilitates SPINE2 application development without dealing with the Java code.

12.3.1   How to Install SPINE2

The process to setup the SPINE2 environment consists of the following steps:

1) Download SPINE2 from the SPINE project website (http://spine.dimes. unical.it/). The folder contains:
   a) *Spine2_ common_c* folder with the sensor-side platform-independent C code.
   b) *Spine2_tinyos-2.x* folder with TinyOS 2.x code to be run on the sensor nodes supporting TinyOS.
   c) *Spine2_coordinator* folder with Java code to be run on the coordinator (i.e. a computer).
   d) *Spine2_console* folder containing the Java code for the GUI.
   e) COPYING and License text files containing info about the licensing.
   f) the SPINE2 manual.
2) The sensor-side folders have to be compiled in TinyOS 2.x and then flashed on sensor nodes. SPINE2 has been developed and tested with TinyOS version 2.1.0. Older TinyOS 2.x versions have also been tested, and the Makefile can be configured to support an older version, but the SPINE team strongly suggests to use TinyOS 2.1.0 release.

a) Copy *spine2_common_c* and *spine2_tinyos-2.x* folders into your tinyos-2.x-contrib folder.

b) From the *spine2_tinyos-2.x/apps/SPINEApp* folder compile and install SPINE2 framework on your TinyOS platform. For instance, if your platform is TelosB:

```
make telosb install,1 bsl,/dev/ttyUSB1
```

where "1" is the sensor node ID (can be freely set by the user) and */dev/ttyUSB1* is the serial port, on Linux machine, to which the sensor node is connected.

3) Configure the TinyOS Java JNI libraries. On Windows machine, copy *toscomm.dll* and *getenv.dll* to *C:\WINDOWS\system32* or to your JRE/JDK *bin* subfolder (i.e. .. \jdk1.xx.xx\bin or .. \jreX\bin). These two files can be found in *$TOSROOT/support/sdk/java/net/tinyos/util* named as *windows_x86_toscomm.lib* and *windows_x86_getenv.lib*. Since these libraries are for 32-bit systems, use a 32-bit i586-JRE version to run the SPINE2 Java apps. On Linux machine, both for 32-bit and 64-bit versions, run in the terminal the following:

```
cd $TOSROOT/support/sdk/java && make
sudo tos-install-jni
```

4) *Spine2_coordinator* and *Spine2_console* can both simply run as any Java application or can be imported as Java project using an IDE such as Eclipse or NetBeans. The console application needs the SPINE2 coordinator project (or its JAR library *SPINE2.jar*) in order to be run. Moreover, for convenience, the *lib* subfolder in both projects already contains the necessary external libraries, like the TinyOS Java library *tinyos.jar*, which can also be found in the *tinyos2.x\support\sdk\java* folder of your TinyOS release. In addition, the Java Communications API is required to support the communication with the sensor node acting as the base station over the serial port. Along with the Java *comm.jar* library, the native binary library needs to be integrated in your operating system:

a) On Windows machines, (i) copy the *win32com.dll* file into the *C:\WINDOWS\system32* folder and (ii) move the *javax.comm.properties* text file to the *lib* subfolder in your JRE folder, i.e. *C:\Program Files\Java\jre6\lib*, by uncommenting the line with the following string: *Driver=com.sun.comm.Win32Driver*.

b) On Linux machines, (i) copy the *libLinuxSerialParallel.so* file into the */usr/lib* folder and (ii) move the *javax.comm.properties* text file to the *lib* subfolder in your JRE folder, uncomment the line with the following string: *driver=com.sun.comm.LinuxDriver*.

12.3.2 How to Use the SPINE2 API

Similar to SPINE 1.x, SPINE2 provides, on the server side, simple Java-API through which a developer can easily develop its own Java application on the coordinator, without dealing with node-side programming issues, for:

- managing the sensor network.
- defining the task-oriented application to be deployed on the WSN.
- managing the (preprocessed) data from the network.

Such a Java application will have to implement the SPINE2Listener interface and the following methods:

`void` `discoveryCompleted(java.util.LinkedList<spine2.wsn.WSNNode>nodes)`
This method is invoked by the SPINE2Manager (through the EventDispatcher) to its registered listeners when the discovery procedure timer fires; it provides a LinkedList of WSNNode objects representing the discovered nodes.

`void` `messageReceived(spine2.message.Message msg)`
This method is invoked by the SPINE2Manager (through the EventDispatcher) to its registered listeners when a new SPINE2 message has been received.

`void` `nodeDiscovered(spine2.wsn.WSNNode node)`
This method is invoked by the SPINE2Manager (through the EventDispatcher) to its registered listeners when it receives a NODE_ADVERTISEMENT_MSG message from a BSN node.

Then, through the SPINE2Manager, whose instance can be only retrieved via the SPINE2Factory, the application can use the following API:

`void` `addListener(SPINE2Listener listener)`
Registers a SPINE2Listener to the manager instance.

`void` `deployApplication(spine2.task.TaskGraph taskgraph, boolean automaticallyStartApp)`
Deploys the task-based application into the sensor network.

`void` `discoveryWSN()`
Commands the SPINE2Manager to discover the surrounding sensor nodes within a timeout (2 s as default).

`long` `getDiscoveryTimeout()`
Gets the discovery procedure timeout.

`spine2.wsn.WSNNode` `getNode(spine2.wsn.Address address)`
Returns a specific sensor node by its address.

spine2.wsn. WSN	**getWSN( )** Returns the object describing the discovered sensor network.
void	**initApplication(boolean automaticallyStartApp)** Initializes the deployed task-based application.
boolean	isStarted() Informs if the task-based application has been started.
void	**removeListener(SPINE2Listener listener)** Removes a SPINE2Listener from the manager instance.
void	**resetApplication()** Removes the task-based application deployed in the sensor network.
void	**startApplication()** Starts the deployed task-based application.

In the following, the API provided by the TaskGraph class for defining the task-based application:

boolean	**addConnection(int sourceTaskCode, int destTaskCode)** Adds a connection to the task graph, by task codes.
boolean	**addConnection(Task sourceTask, Task destTask)** Adds a connection to the task graph.
boolean	**addConnections(Task sourceTask, Task[] destTasks)** Adds a set of connections to the task graph, from one source task to multiple destination tasks.
boolean	**addTask(Task task)** Adds a task instance to the task graph.
boolean	**connectionAlreadyExist(int sourceTaskCode, int destTaskCode)** Verifies if a connection has been instantiated.
java.util. LinkedList <Connection>	**getAllInputConnections(int taskCode)** Returns the list of input connections connected to a specific task.
java.util. LinkedList <Connection>	**getAllOutputConnections(int taskCode)** Returns the list of output connections given a specific task.
Connection	**getConnection(int sourceTaskCode, int destTaskCode)** Returns the connection between two tasks given their codes.

java.util. LinkedList <Connection>	`getConnectionsList()` Returns the list of all connections.
Task	`getTask(int taskCode)` Returns a specific task instance given its code.
Task	`getTask(java.lang.String logicalName)` Returns a specific task instance from its logical name.
java.util. LinkedList <Task>	`getTaskList()` Returns the list of tasks in the task graph.
boolean	`removeConnection(Task sourceTask, Task destTask)` Removes a connection from the task graph.
boolean	`removeTask(Task task)` Removes a task instance from the task graph.
void	`reset()` Resets the application, i.e. all its related information (tasks and connections) is deleted.
boolean	`updateTask(Task task)` Updates a task instance already into the task graph.

12.3.3 How to Run a Simple Application Using SPINE2

The SPINE2 release comes with a simple test application (SPINE2SimpleTest. java) that can be run to experiment the framework basic functionalities. Assuming the use of the TinyOS environment on the sensor nodes, running the application consists in the following steps:

1) Compile and flash, on an available sensor node, the SPINE2 TinyOS node-side software.
2) Compile and flash a TinyOS2.x BaseStation onto another sensor node. It is important to check that sensor nodes and base station are both working on the same radio channel, have been compiled with the same max message payload length, and the same TinyOS version has been used for flashing all the nodes.
3) Plug the BaseStation to a free USB port of the computer and type *motelist* from your shell: this will return the USB port number.
4) Create an application properties file (e.g. under *MyApp/resources/myapp. properties*) and set the "enabled_platforms" and the "platform_motecom" parameters according to one of the following options, depending if the base station is communicating with the USB serial port on a Linux (a) or Windows machine (b):

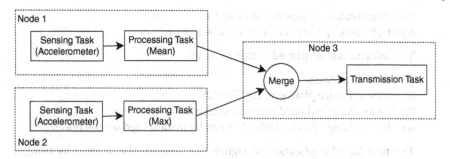

Figure 12.1 The task-oriented application defined and deployed in "SPINE2SimpleTest.java" application.

a) `enabled_platforms=tinyos`
   `tinyos_motecom=serial@/dev/ttyUSB0:telosb`
b) `enabled_platforms=tinyos`
   `tinyos_motecom=serial@COM0:telosb`

SPINE2 currently fully supports TinyOS platforms, whereas support for Z-Stack is under development.

5) Edit the *Spine2_coordinator/test/SPINE2SimpleTest.java* and go through the code if you want to customize the test application.

The *SPINE2SimpleTest* application in the current SPINE2 release has been created to show how to develop the task-based application of Figure 12.1 over a BSN composed of three sensor nodes.

As mentioned before, SPINE2SimpleTest.java implements the SPINE2-Listener interface (to get notified of SPINE2-related events and messages) and uses the SPINE2Factory to retrieve the SPINE2Manager, which, in turn, has the APIs for managing and communicating with the nodes in the network. Moreover, a TaskGraph instance has been created and modeled to reflect the user-defined task-based application.

In particular, the sample application performs the following actions:

a) A discovery message is broadcast to discover the surrounding SPINE2-capable sensor nodes, after which the manager collects the reply messages sent by the nodes within `DISCOVERY_TIMEOUT=3000`ms.

```
manager.setDiscoveryTimeout(DISCOVERY_TIMEOUT);
manager.discoveryWsn();
```

b) Once the discovery is completed, the application is notified by the manager through the `discoveryCompleted(LinkedList<WSNNode>nodes)` method, which returns the list of discovered sensor nodes, whose information is then displayed.

```
currentNode = motes.get(j);
System.out.println(currentNode);
```

The information displayed at this point is:

i) node id/address.
ii) node software platform, e.g. TinyOS.
iii) the available onboard physical sensors.
iv) the available SPINE2 tasks that can be instantiated on the node.

The task-based application of Figure 12.1 can now be defined by using a `TaskGraph` instance.

```
taskGraph= new TaskGraph();
```

The tasks to be instantiated on the first node are then created and added to the `TaskGraph` instance. The sensing task is configured to periodically acquire sensed data from the onboard accelerometer with sampling time `SAMPLING_TIME=50` ms. The processing task is configured to compute the mean over the accelerometer raw data with `WINDOW_SIZE=40` and `SHIFT_SIZE=20`.

```
//      SENSING TASK
SensingTask sensingTask1= new SensingTask(motes.
get(0));
sensingTask1.setLogicalName("Sensing_Task_1");
sensingTask1.setSensorType(Sensor.ACCELEROMETER);
sensingTask1.setPeriodicity(SensingTask.TIMER_PERIODIC);
sensingTask1.setTimeScale(SensingTask.TS_MILLISEC);
sensingTask1.setPeriod(SAMPLING_TIME);
sensingTask1.setDataSelection(SensingTask.DATA_ALL);
taskGraph.addTask(sensingTask1);
//      PROCESSING TASK
ProcessingTask procTaskMean1=
                        new ProcessingTask(motes.
                        get(0));
procTaskMean1.setLogicalName("Processing_Mean_1");
procTaskMean1.setFunctionType(FunctionConstants.F_MEAN);
procTaskMean1.setWindowSize(WINDOW_SIZE);
procTaskMean1.setShiftSize(SHIFT_SIZE);
procTaskMean1
      .setOutputBuffering(PROCESSING_OUTPUT_
      BUFFERING);
taskGraph.addTask(procTaskMean1);
```

c) A similar configuration is defined for the second node.

```
//      SENSING TASK
SensingTask sensingTask2= new SensingTask(motes.get(1));
sensingTask1.setLogicalName("Sensing_Task_2");
sensingTask1.setSensorType(Sensor.ACCELEROMETER);
sensingTask1.setPeriodicity(SensingTask.
TIMER_PERIODIC);
sensingTask1.setTimeScale(SensingTask.TS_MILLISEC);
sensingTask1.setPeriod(SAMPLING_TIME);
sensingTask1.setDataSelection(SensingTask.DATA_ALL);
taskGraph.addTask(sensingTask2);
//      PROCESSING TASK
ProcessingTask procTaskMean2=
                    new ProcessingTask(motes.
                    get(1));
procTaskMean1.setLogicalName("Processing_Mean_2");
procTaskMean1.setFunctionType(FunctionConstants.F_MEAN);
procTaskMean1.setWindowSize(WINDOW_SIZE);
procTaskMean1.setShiftSize(SHIFT_SIZE);
procTaskMean1
      .setOutputBuffering(PROCESSING_OUTPUT_
      BUFFERING);
taskGraph.addTask(procTaskMean2);
```

d) Finally, the tasks for the third node.

```
//      MERGE TASK
MergeTask mergeTask= new MergeTask(motes.get(2));
mergeTask.setLogicalName("Merge_Task");
taskGraph.addTask(mergeTask);
//      TRASMISSION TASK
TransmissionTask transmTask =
                    new TransmissionTask(motes.
                    get(2));
transmTask.setLogicalName("Transmission_Task");
transmTask.setDestinationAddr(
              CommConstants.SPINE_BASE_STATION_
              ADDR);
taskGraph.addTask(transmTask);
```

e) Next, the connections between pair of tasks are created.

```
taskGraph.addConnection(sensingTask1,
procTaskMean1);
taskGraph.addConnection(sensingTask2, procTaskMean2);
taskGraph.addConnection(procTaskMean1, mergeTask);
taskGraph.addConnection(procTaskMean2, mergeTask);
taskGraph.addConnection(mergeTask, transmTask);
```

f) Once the application task graph is defined, it can be deployed over the network. Moreover, the manager is instructed to automatically run the task application as soon as all the tasks are instantiated on the sensor nodes.

```
manager.deployApplication(taskGraph,
             WSN.AUTOMATICALLY_START_APPLICATION);
```

g) As last operation of the discoveryCompleted(...) method, the MetaDataManager instance is used to build the metadata information related to the just-defined task graph application. This component is necessary to correctly extract the sensor data from the SensorDataMessage.

h) On the reception of a message from the sensor network, and specifically from the TransmissionTask instance, the messageReceived (Message msg) method is triggered in order to handle such a message on the basis on its type (see "spine2.message.message_list" package). In particular, in case of a SensorDataMessage, the Meta DataManager instance is used to allow the developer to simply extract the data of interest, which can be identified by means of specific labels. Moreover, the check over the dataMsgChainID value can be useful to differentiate data messages coming from different transmission tasks, which is not actually necessary in this case. Data are then simply displayed with no further computation.

```
if(msg instanceof SensorDataMessage){
    SensorDataMessage dataMsg=
             (SensorDataMessage) msg;
    metaDataManager.decodeSensorDataMsg
(dataMsg);
    int dataMsgChainID= dataMsg
             .getTransmissionTaskCode();
    if(dataMsgChainID== transmTask.getCode()){
         short[]streamMeanX= metaDataManager
             .getDataStream(
             "Mean_AccX_Sensing_Task_1");
```

```
if(streamMeanX!=null)
        printDataStream(
                    "Mean_AccX_Sensing_
                    Task_1",
                    streamMeanX);
else
        System.out.println("No data
        associated
                    with the specified
                    label");
    }
    }
```

i) Finally, in order for the developer to know the exact list of available sensor data labels, the `MetaDataManager` provides the following method:

```
MetaDataManager.getMetaDataLabelsListString(
                    TaskGraph
                    taskgraph);
```

Index

Wearable Computing: From Modeling to Implementation of Wearable Systems Based on Body Sensor Networks, First Edition. Giancarlo Fortino, Raffaele Gravina, and Stefano Galzarano.
© 2018 John Wiley & Sons, Inc. Published 2018 by John Wiley & Son, Inc.